PRENTICE HALL

Statistics

MINITAB Manual
PRENTICE HALL

Statistics

Ivy Lee, Ph.D

Minako K. Maykovich, Ph.D
California State University

Prentice-Hall Inc., Englewood Cliffs, New Jersey 07632

Editorial production/supervision: **Sandra Lynn Barrett**
Manufacturing buyers: **Paula Massenaro/Lori Bulwin**
Acquisitions editor: **Steven Conmy**
Supplement acquisitions editor: **Alison Munoz**

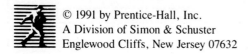

© 1991 by Prentice-Hall, Inc.
A Division of Simon & Schuster
Englewood Cliffs, New Jersey 07632

All rights reserved. No part of this book may be
reproduced, in any form or by any means,
without permission in writing from the publisher.

Printed in the United States of America

10 9 8 7 6 5 4 3 2 1

ISBN 0-13-852971-X

Prentice-Hall International (UK) Limited, *London*
Prentice-Hall of Australia Pty. Limited, *Sydney*
Prentice-Hall Canada Inc., *Toronto*
Prentice-Hall Hispanoamericana, S.A., *Mexico*
Prentice-Hall of India Private Limited, *New Delhi*
Prentice-Hall of Japan, Inc., *Tokyo*
Simon & Schuster Asia Pte. Ltd., *Singapore*
Editora Prentice-Hall do Brasil, Ltda., *Rio de Janeiro*

TABLE OF CONTENTS

CHAPTER ONE	INTRODUCTION	1
1.1	Data on which Minitab works	1
1.2	The Minitab worksheet	2
1.3	Minitab command language and syntax	3
1.4	Minitab operating modes	5
1.5	Running a simple Minitab session	6
1.6	Correcting errors	9
1.7	Getting help	10
1.8	Quitting a Minitab session	10
1.9	Organization of this book	11
1.10	More on commands	11

CHAPTER TWO	DATA ENTRY AND TABULATION	17
2.1	Data entry utility: Data Editor	17
2.2	Saving and retrieving the data	25
2.3	Column and row manipulation commands	27
2.4	Tabulating data	29
2.5	More on commands	34

CHAPTER THREE	DESCRIPTIVE STATISTICS	40
3.1	Arithmetic operations	40
3.2	Summarizing data in columns or rows	45
3.3	Summarizing data in cells	48
3.4	Comprehensive summary of data	49
3.5	Column and row manipulation	50

CHAPTER FOUR	GRAPHING DATA	58
4.1	The Minitab and DOS operating system	58
4.2	Univariate Plotting: One or K sample data	64
4.3	Bivariate Plotting	71
4.4	Multivariate Plotting	74

CHAPTER FIVE	PROBABILITY AND SAMPLING DISTRIBUTION	81
5.1	Concept of probability	81
5.2	Probability of events	84
5.3	Probability distributions	86
5.4	Sampling distributions	92
5.5	The Central Limit Theorem	95
5.6	More on Commands	100

CHAPTER SIX	TESTS OF MEANS	104
6.1	One Sample Mean	104
6.2	Two Sample Means	111

CHAPTER SEVEN	ANALYSIS OF VARIANCE	119
7.1	Oneway analysis of variance	119
7.2	Randomized block design	124
7.3	Twoway analysis of variance: Balanced Data	127
7.4	Unbalanced data	133

CHAPTER EIGHT	NONPARAMETRIC TESTS	138
8.1	One-sample test	138
8.2	Two sample test	144
8.3	K-sample test	146
8.4	Bivariate Relations	151

CHAPTER NINE	CORRELATION AND REGRESSION	158
9.1	Correlation coefficient	158
9.2	Regression coefficient	161
9.3	Confidence and prediction intervals	167

CHAPTER TEN	MULTIPLE REGRESSION	172
10.1	Building a regression model	172
10.2	Examining the nature of the data	175
10.3	Simplifying the model: Stepwise regression	177
10.4	Simplifying the model: BREG	179
10.5	Polynomial regression	181
10.6	Models with dummy variables	183
10.7	More on commands	187

INDEX   190

PREFACE

This is a Minitab Guide to accompany any general statistics textbook. It provides an introduction to the PC Version 7.2 of Minitab, which is a powerful, yet easy-to-use statistical package. It also prepares the student to use the mainframe version of Minitab as the differences between the two are minor.

To benefit from the advances in computer technology, more and more statistics courses are taught with statistical packages such as Minitab. Not only can the computer aid in making calculations easier and faster but also in enhancing the student's understanding of statistics. For example, simulating the process of sampling to illustrate the Central Limit Theorem is impossible without the aid of a computer. By experimenting with a large number of samples generated by Minitab, the student comes to grips with the probability theory behind the statistical techniques more readily than by simply reading a textbook.

Among various statistical packages such as SAS, SPSS, or BMDP, Minitab is the easiest to approach, although these packages all accomplish basic statistical tasks. In the PC version of Minitab, the student communicates interactively with the computer in such a way that the machine will respond to commands instantaneously. Without learning a great deal about the setups and the command syntax, the student can begin doing statistical analysis during the first Minitab session.

Ease of initiation is not the only merit of Minitab. Minitab performs very complex statistical analyses on large scale data. As such, it is used not only in academia, but also in business, industry, government, and other institutions.

There are many Minitab manuals already in print. Some are elementary treatments of statistics, while others are encyclopedic references of commands syntax. Still others are supplements to certain textbooks and cannot be used independently.

What is unique about this book is that, within limited space, it introduces Minitab techniques gradually and progressively. Beginners may stop at any point within a chapter or within the manual if they are satisfied with their accomplishments. On the other hand, advanced students may delve into the more complex chapters with their exposition of sophisticated analyses, thus making the most of available Minitab facilities.

This Guide is written in such manner as to require a minimum of computer knowledge and experience. A basic knowledge of DOS would be helpful, although not necessary. In addition, the data used within the chapters are not tied to any specific textbook; hence the book can be used by students of statistics in various fields.

Minako K. Maykovich
Ivy Lee

PRENTICE HALL

Statistics

CHAPTER ONE
INTRODUCTION

```
In This Chapter You Will Learn To:

  1.  Organize and transform data for Minitab analysis
  2.  Obtain an overview of the Minitab worksheet, command
      syntax, and operating mode
  3.  Run a simple Minitab session for data input and
      output
```

The purpose of this book is to introduce you to the Minitab Statistical Package for Personal Computers. Minitab is a set of prewritten computer programs which can be used for many types of statistical analyses. It is easy to use even though it handles very complex statistical procedures. Minitab on Personal Computers can be used interactively so that you and the computer can communicate with each other instantaneously. After each command you enter on the screen, you receive feedback from the machine with very little waiting time in between. Furthermore, an on-line HELP facility is available, which provides a short explanation for each command and its syntax. Minitab is an invaluable tool for a data analyst.

1.1 DATA ON WHICH MINITAB WORKS

Before discussing how Minitab works, consider how data must be prepared so that Minitab can process them. The data collected by a researcher must be transformed before the computer can read and make sense of them.

The following hypothetical dataset of a class will be used to illustrate concepts regarding data collection and transformation. For each student of the class the scores for two mid-term exams and the final exam are shown below together with the student's name and major:

Name	Major	Exam 1	Exam 2	Exam 3
Abbot,John	1	90	95	90
Baker,Kathy	2	100	92	89
Cook,Dianne	3	80	80	82
Hayes,Fred	1	65	60	60
	2	89	*	78

The dataset contains 5 variables: the name, major, the first and second exam scores, and the final exam score. Variables are so called because they vary, i.e., take on different values or attributes for each observation or student. Very generally, Minitab classifies variables into two types: alphabetical and numerical.

There is one alphabetical variable in the dataset: name. Names are entered in alpha codes, which can consist of combinations of letters of the alphabet, special characters and numbers. The rest of the variables are numerical. Numerical variables can be represented by the actual numerical values or values which symbolize categorical classifications. In our example, the scores of Exam 1, Exam 2, and Final are entered directly as raw data, but major is represented by numerical categories as follows:

1	for	Science
2	for	Social Science
3	for	Humanities
4	for	Others

Numerical variables such as Exam 1, Exam 2, and Final, are expressed either as integers or in decimals. Special characters such as the comma (,) and the dollar sign ($) must be dropped. Thus the price of a house, $185,950 must be entered as 185950. A stock price of $15 3/8 is transformed into 15.375.

If a numerical value is <u>missing</u>, enter an asterisk (*), such as shown in the last row of the dataset illustration. Missing values in alpha columns (columns with alphabetical rather than numeric data) are denoted by blank spaces, as demonstrated in the same illustration in the last row below the name "Hayes".

Whether characters or numbers are used, they are referred to as codes for the values of a variable. The whole process of transforming data into these codes which are readable by the computer is known as <u>coding</u>.

1.2 THE MINITAB WORKSHEET

Minitab stores data in (1) <u>columns</u> (C1, C2, C3, ...), each of which contains one or more numbers, (2) boxes of <u>constants</u> (K1, K2, K3, ...), each storing one number, and (3) matrices (M1, M2, M3, ...).

When data are stored in columns, they are entered in the form of a worksheet, which is a rectangular array of numbers formed by vertical columns and horizontal rows. Each column contains the values or codes of a different variable. In Minitab notation columns are referred to as C1, C2, and C3 for the first, second, and third columns. Thus C1 may refer to a Student's name, C2 is the Major, C3 Exam 1, C4 Exam 2, and C5 the Final exam.

Each row of the worksheet refers to a particular observation, case or student. In the example cited, the first row contains the data for John Abbot, the second row, for Kathy Baker, and so on.

In addition to the rectangular array, Minitab provides locations where constants can be stored, denoted by K1 for the first constant, K2 for the second, and so on.

Matrices are denoted by M1, M2, ... Each M stores a matrix. Most of the time, your data are in the form of columns and constants rather than matrices. Unless otherwise indicated, discussions of the input and output procedures in this chapter therefore pertain to columns and not matrices.

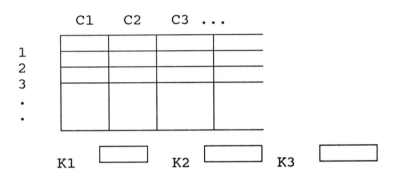

Columns, stored constants, and matrices can be used again and again. When they are reused, new values may supersede previous contents.

The storage capacity of Minitab, that is, the total number of columns, stored constants, and matrices entered and available for statistical manipulation, depends on the type of computer used. Use the HELP facility (HELP OVERVIEW 8) to obtain information on storage for your computer.

1.3 MINITAB COMMAND LANGUAGE AND SYNTAX

A command tells Minitab which task is to be performed. Commands begin with command names, sometimes referred to as command keywords, examples of which are PRINt, READ, SET, and REGRess. Note that the first four letters are capitalized here because Minitab recognizes only these letters. After the first four letters of a command, the rest is superfluous; they may be entered or left out. The command names consisting of the first four letters are stored in a dictionary which Minitab uses to check on the command everytime it is called forth. Although a distinction is made here for the ease of communication, you may type a command name in either upper or lower case.

The command name or keyword is often followed by argument(s). An argument can be a column, a stored constant, a matrix, a number, or a file name. For example, to print the contents of column 3, the following command keyword and arguments are necessary:

```
MTB > PRINt C3
```

The whole line, command name plus argument is referred to as a command.

Each command must begin on a new line. If a command is too long, type an '&' as the last character on the first line, and continue on to a second line. All commands and data are in free format, which means that they may be placed anywhere on the line. It is not the position, but the order of entry which counts.

Some commands are accompanied by <u>subcommands</u> to provide options or produce additional output. If subcommands are used, the main command appears on a line by itself, followed by a semicolon. The subcommands follow each beginning in a new line and ending in a semicolon. The last line must end with a period. The period is unnecessary for commands without subcommands.

```
MTB>  REGRess 'Normal' 1 'Over WGT' C20 C21;
SUBC> COEFficients C22;
SUBC> RESIduals C23.
```

Do not worry about the meaning of the above example. Simply be able to identify the components of the command and subcommand language. REGRess is the command where the first four letters REGR are recognized by the Minitab system. 'Normal' 1 'Over WGT' C20 C21 are arguments specifying the variables needed for the execution of the command. The command line is terminated with a semicolon because it is followed by subcommands. COEFFicients is the subcommand followed by its arguments, C22. After this first subcommand ends with a semicolon, the second subcommand, RESIduals, appears with its argument, C23 on a new line. Since this is the last subcommand, it terminates with a period.

You can add extra text for annotation in all commands except LET, ANOVA or ANCOVA. However, do not use numbers or the symbols ' . & + - # : for annotation. The following pair of commands accomplish the same task:

```
MTB > MEAN of C1, put in K1
MTB > MEAN C1 K1

MTB > REGRESS income in C1 on 1 education in C2
MTB > regr c1 1 c2
```

If you made a mistake in typing a subcommand, type ABORT as the next command. This cancels the entire command, allowing you to start again from the main command.

Throughout this book command syntax is presented in a box. Notations for arguments are as follows:

K   Denote a constant which can be a number or a stored constant
C   Denote a column
E   Denote either a number, stored constant or a column
M   Denote a matrix
[ ]   Denote an optional argument

1.4 MINITAB OPERATING MODES

Minitab commands can be executed in two modes: batch and interactive. In addition, there is the Data Editor which makes the data entry process efficient.

In the <u>batch</u> mode you write a program on a terminal using an editor (such as EDLN in DOS), submit the entire program to the computer, wait for a while, and then receive all the results at once. Do not confuse an editor, which is not an integral part of the Minitab package with the Data Editor which comes with Minitab. An editor allows you to change and correct computer entry on screen among other things. The Data Editor of Minitab has a much more limited and specific function. It is designed for entering and changing data only. Detailed explanation of the use of the Data Editor will be given in the next chapter.

When Minitab works <u>interactively</u>, you send one command at a time, and the computer responds immediately before you send another message. Minitab reads the command, looks it up in its dictionary, and checks for errors. If you have made a mistake in syntax or typing, the computer returns an error message. When the command is accepted, a prompt will appear on the screen, directing you to proceed. If the command requests some action such as printing the results or computing statistics, the action will take place first, then a prompt appears afterwards.

Minitab prompts which appear in the interactive mode are shown below:

MTB>   Command prompt

SUBC>   Subcommand prompt, which appears following a semicolon of the previous line

DATA>   Data prompt, which appears after the commands, READ, SET and INSERT

CONT>   Continuation prompt, which appears when the previous line ends with an '&'. It allows a command, subcommand or data row to be entered in more than one line.

CONTINUE?   Paging command which appears when Minitab pauses in the process of output or display. Press ENTER to see more output. Otherwise type N or No, then ENTER.

1.5 RUNNING A SIMPLE MINITAB SESSION

1.5.1 Entering Data One Column at a Time

1.   Type MINITAB in the appropriate subdirectory, and press ENTER to invoke Minitab. The MTB> prompt appears on the screen, waiting for your command.

The SET Command

2.   To enter data one column at a time, we use the SET command followed by the column specification. When so doing, we are recording information on one variable over all observations. For example, we can enter students' names from our dataset into column one. The SET command automatically assumes that numerical data will be entered. Since student names are alpha data, the FORMAT subcommand is needed for specification in addition to the SET command. Syntax of this command is explained at the end of this chapter.

```
MTB > SET names into C1;
SUBC> FORMAT(A20).
DATA> Abbot,John
DATA> Baker,Kathy
DATA> Cook,Dianne
DATA> Hayes,Fred
DATA>
DATA> END
MTB >
```

At the MTB> prompt, type the command keyword, SET, followed by the column specification, C1. As explained previously, Minitab responds only to the keyword SET and C1, ignoring commentaries such as "names into".

Since the FORMAT subcommand is needed in alpha data entry, finish the first command line with a semicolon. Upon pressing ENTER, the SUBC> prompt appears so that you can enter a format statement. After typing the keyword, FORMAT, enter a format specification in parentheses. "A" stands for alpha data, and "20" specifies the width of the alpha field. In other words, this column can handle a student's name which is as long as 20 characters. Finish this subcommand with a period. The period tells the computer that it is the end of the command; no other subcommand can be expected.

Upon pressing ENTER, the DATA> prompt appears for data entry. This prompt follows data entry commands only and indicates that the computer is expecting you to input data. You must enter one name per line, and press ENTER after each entry. Note that the last name is missing. Therefore, do not enter anything after the DATA> prompt for that line, but simply press ENTER.

When you complete the data entry, type END, and press ENTER. The MTB> prompt appears for the next command. The END command informs Minitab that data entry is completed. When entering data from the keyboard, it is advisable to use this (optional) END command, following the last line of data.

Displaying the Contents on the Screen

3. In order to see if the data are entered correctly, use the PRINT command to display the contents of the column(s) specified after the command name.

```
MTB > PRINT C1
 C1
    Abbot,John          Baker,Kathy          Cook,Dianne
    Hayes,Fred
```

Entering Numerical Values by SET

4. Numerical data can be entered without a format specification. Let's enter the students' majors into column 2.

```
MTB > SET majors into C2
DATA> 1 2 3 1 2
DATA> END
MTB >
```

Since there is no subcommand for SET, there is neither a semicolon nor a period at the end of the first line. With numeric data, you have a number of options in entering them. You can type one value per data line as you did when entering the alpha data. Or you can enter all values for C2 in one line as shown in the example above, or break them into several DATA lines with several values per line. The numbers on the data lines are delimited by space and/or commas. After completing data entry, do not forget to type END, and press ENTER.

5. Let us view the contents of C1 and C2 by the PRINT command. The two columns can be specified as: [C1 C2], [C1,C2], or [C1-C2]

```
MTB > PRINT C1 C2
  ROW                          C1        C2
      1      Abbot,John                   1
      2      Baker,Kathy                  2
      3      Cook,Dianne                  3
      4      Hayes,Fred                   1
      5                                   2
```

7

1.5.2 Entering Data Several Columns at a Time

In most situations when the data consist of more than one variable you will wish to input data into several columns at once instead of dealing with each column separately. In such cases, data are entered row-wise into the worksheet. Let us finish entering the exam scores into columns 3 through 5 from our class dataset. When inputting data several columns at a time or row-wise, the SET command is inappropriate. The READ command is used instead.

1. The READ command below specifies that data should be entered into columns 3 through 5. Note that since there are three columns, the dash '-' must be used to indicate that you are reading into columns 3 through 5. Otherwise, if you specify [C3 C5], only two columns are set up for use. Following the DATA> prompt, enter the numbers of a row separated by space and/or commas. If the numbers for three columns are to be broken into more than one data line as shown in line 3 at the DATA> prompt in the following example, use the '&' symbol. In this example, two numbers (80 80) are entered followed by '&'. Minitab provides the CONT> prompt to let you enter the third number (82) in the following line. Do not enter more values than the number of columns specified in the READ command; otherwise you will get an ERROR message.

    ```
    MTB>    READ C3-C5
    DATA>   90 95 90
    DATA>   100,92, 89
    DATA>   80, 80 &
    CONT>   82
    DATA>   65 60 60
    DATA>   89  *  78
    DATA>   END
    ```

2. Using the PRINT command, let us verify the data entry.

    ```
    MTB > PRINT C1-C5
      ROW                       C1   C2    C3    C4    C5
         1    Abbot,John              1    90    95    90
         2    Baker,Kathy             2   100    92    89
         3    Cook,Dianne             3    80    80    82
         4    Hayes,Fred              1    65    60    60
         5                            2    89     *    78
    ```

    As you can see from the screen display, everything is in order. Columns 3 through 5 are added onto columns 1 and 2. The worksheet now comprises five columns.

1.5.3 Naming Variables

1. The NAME command assigns a descriptive name to a column. The name can be up to eight characters long and must be enclosed in single quotation marks. After a

name is given, reference to a column can then be made either by its number or name in the Minitab command. Let us attach names to the columns in the class data.

```
MTB>   NAME C1 'Name' C2 'Major' C3 'Exam1' C4 'Exam2' &
CONT> C5 'Final'
```

2. The PRINT command displays the data with column names. Since names have been assigned, you can specify the columns by column numbers or column names in the PRINT command. Regardless of how you specify the columns after the PRINT, the output will now display the column names in place of C1, C2...C5.

```
MTB > PRINT C1-C5
or
MTB > PRINT 'Name' 'Major' 'Exam1' 'Exam2' 'Final'
```

1.5.4 Placing Comments

1. It is a good idea to document your work by placing comments where appropriate. It helps you and others to follow what you are doing. The NOTE command or the # symbol inserts comments as follows:

```
MTB > NOTE Statistics 101
MTB > NOTE Major: 1=Science   2=Social Science   3=Humanity &
SUBC> 4=Other
MTB > #  List majors below
```

2. A # symbol can be used in the same line where a command lies. You can then write comments after a # symbol, since Minitab ignores everything between # and the end of a line. For example:

```
MTB>   PRINT C3-C5    # Print exam scores
```

1.6 CORRECTING ERRORS

In general when you made a typing error while you are still on the same line, you can use the BACKSPACE key to move to the error location, and correct it. Backspacing will erase entry back to the point where the error is made. In addition, the left arrow key performs the same function.

Specifically, for data entry, if an error is discovered after entering the entire data set and printing the result, you can use the LET command to change data values. The LET command is very versatile; its other uses will be discussed in later chapters. Suppose the true score of the final exam (C5) for the second person (Row 2) is 100 instead of 89. You can make the correction by reassigning a value. Since C5 has been named as 'Final', the LET command can be written as:

```
MTB>   LET C5(2)=100
or
MTB>   LET 'Final'(2)=100
```

After the command is executed, the value of the fifth column, second row is changed to 100. Since LET alone does not produce any output, you must use PRINT to confirm that the value has really been changed. This process of correcting errors is quite tedious if there are a number of errors. For more efficient editing, Minitab provides the Data Editor utility, which will be discussed in Chapter 2. Also, for more extensive editing, there are INSERT, DELETE, and ERASE commands to remove rows, columns, or a rectangular portion of the data. These commands are presented also in Chapter 2.

1.7 GETTING HELP

To help you understand the general structure of Minitab and obtain a summary explanation for each command and subcommand, Minitab offers an excellent Help facility. Whenever you are in doubt, type HELP at the prompt. To learn about Minitab, type: HELP HELP, and follow the instructions on the screen. Thus:

HELP   Display a general introduction to the HELP facility

HELP HELP   Provide Help facility to help you learn about Minitab

Other help facilities are also available, such as:

HELP OVERVIEW   Provide general help

HELP COMMANDS   Provide a list of all HELP topics available.

HELP SET   Provide information about using a specified command, such as
                          SET

1.8 QUITTING A MINITAB SESSION

1.8.1 Leaving Minitab at the End of a Job

To end your Minitab session, issue the command STOP. Upon pressing ENTER, you will be returned to the DOS operating system. Since ending the session will erase the current worksheet, be sure to save any data you will need in the future. Data can be saved using either the WRITE or the SAVE command. These commands are discussed in Chapter 2

```
MTB > STOP           # Press ENTER, then you will be back to DOS
```

1.8.2 Terminating the Execution of a Minitab Program

Sometimes it is necessary to terminate a lengthy execution of a Minitab program before it is completed. In such cases, press <u>CTRL-BREAK</u>, to return you to the DOS environment.

1.9 ORGANIZATION OF THIS BOOK

This book is divided into basic statistics topics so that a reader can use it together with any general introductory statistics textbook.

The first two chapters provide an introduction to Minitab, giving an overview of the tools with which Minitab operates. In this chapter, you have learned to run a simple Minitab session interactively. In Chapter 2 the Data Editor utility is introduced to make data entry and editing easy and fast.

The remaining chapters are geared largely to statistical analysis, presenting Minitab statistical commands and subcommands. Along with these commands, various operating modes are introduced gradually and progressively. They include among other things: interactive versus batch mode, keyboard data entry versus data file input, and the macro facility. As you become familiar with the Minitab language, the focus shifts from writing to interpreting programs.

Each chapter begins with a summary of what you will learn in that chapter, and closes with a list of new Minitab commands covered. Within most chapters, a statistical problem is first posed, then a Minitab program follows. After executing the Minitab commands, the outputs are displayed together with interpretation. To streamline the presentation, detailed explanations of commands, if necessary, are presented at the end of the chapter.

By the end of the book you will have a fair grasp of Minitab commands and operations. An Index of Commands at the end of the book together with a list of the new commands in each chapter will help you look up the commands you need.

1.10 MORE ON COMMANDS

1.10.1 The SET Command

Syntax:
```
SET   [from 'filename'] into C

      FORMAT (format specification)
      NOBS = K
```

11

The SET command enters data into one column. If you specify a filename, data are read from the file. Otherwise you must enter data from the keyboard at the DATA prompt, following the SET command.

Normally data are entered individually and in a free format, that is, without the FORMAT subcommand, separated by blank(s) or comma(s). To enter a range of integers from 1 to 50 into column nine, use a colon between lower and upper bounds as follows. This way, you eliminate the tedious task of typing fifty values.

```
MTB > SET C9
DATA> 1:50
```

If the data values increase by other than 1, then the increment may be given after a slash. The following example enters the values 1.0 1.5 2.0 2.5 3.0 3.5 4.0.

```
MTB > SET C10
DATA> 1:4/0.5
```

To enter many identical numbers or sequences of numbers, use parentheses. A repeat factor, with no space in between, may be put in front of the parentheses, after the parentheses, or both before and after. If the repeat factor, K, is in front, the entire list within the parentheses is repeated K times. If the repeat factor K is last, each number is repeated K times before going on to the next number. If both are specified, the last repeat factor is executed first, then the front repeat factor.

```
MTB > SET INTO C11
DATA> 3(1 2 3)
```

By this command, C11 will contain the numbers: 1 2 3 1 2 3 1 2 3

```
MTB > SET INTO C12
DATA> 3(1 2 3)2
```

The above command enters the following numbers into column 12:
1 1 2 2 3 3 1 1 2 2 3 3 1 1 2 2 3 3

The FORMAT subcommand specifies in which way the data are to be read or to be printed. It is used when:

(1)   alpha data are input.
(2)   input data are not delimited by space or a comma.
(3)   the number of decimal digits in numbers are to be specified.
(4)   data are to be output in a format other than the default.

12

Acceptable formats to be enclosed in parentheses are as follows:

Fw.d
For numbers with or without decimal digits
w is the width of the total field including the decimal point, and d is the number of decimal digits within the field. Fw.d reads a number from the next w spaces, and inserts a decimal point d spaces from the right of that number, if a decimal point is not already inserted. For example:

(F5.2, F3.0) will read the data 12345678 as 123.45 and 678.

Aw
For alpha data
w refers to the width of the field where a character string is to be identified. The field width may be up to 80 with SET, READ, or INSERT. Columns read with an A format can be output by WRITE and PRINT with or without a FORMAT command. It can be converted to numbers with CONVERT.

Ew.d   For exponential format
nX   Skip n spaces
Tn   Move to position n
n   Repeat an operation n times
,   Separate format statements
/   Skip a line

Missing data may be indicated by a single * in the numerical field or by leaving the alpha field blank.

The NOBS subcommand specifies the number of observations to be entered. It is especially useful when you want only the first portion of a file read.

1.10.2 The READ Command

```
Syntax:     READ  [from 'filename'] into C,...,C

            FORMAT (format specification)
            NOBS = K
```

The READ command enables you to enter data several columns at a time. If a file is specified, data are read from the file. Otherwise, columns of data are entered row-wise from the keyboard at the DATA prompt, following the READ command.

Without a FORMAT specification, data are entered in free format, separated by space(s) or comma(s). For details about the FORMAT subcommand, see the SET command syntax.

In the earlier example if you wish to enter class data for five columns all at once, you will need a format subcommand because there are alpha data involved. The FORMAT subcommand specifies that C1 contains alpha data of 20 letters in length. "2X" means skipping two spaces. C2 has one-digit (F1.0) numerical data. C3, C4, and C5 have three-digit numerical data (F3.0), separated by two spaces. "3" before a parenthesis is a multiplier, indicating that (2X,F3.0) will be repeated three times. It is the same as (2X, F3.0, 2X, F3.0, 2X, F3.0).

```
MTB > READ C1-C5;
SUBC> FORMAT(A20, 2X, F1.0, 3(2X, F3.0)).
DATA> Abbot,John        1   90   95   90
DATA> Baker,Kathy       2  100   92   89
DATA> Cook,Dianne       3   80   80   82
DATA> Hayes,Fred        1   65   60   60
DATA>                   2   89    *   78
DATA> END
```

1.10.3 The NAME Command

The NAME command is used both for naming initially or renaming a column or columns. The need for renaming arises when in a continuous session you wish to reuse the same columns for other variables with or without labels. In the following example, the names of the columns, C1-C2, are blanked out by the NAME command in the renaming process. Note that there is no space between two single quotation marks to indicate blanking.

```
MTB > READ C1 C2
DATA> 100 95
DATA> END
MTB > NAME C1 'Exam1' C2 'Exam2'
MTB > PRINT C1-C2
 ROW      Exam1       Exam2
   1        100          95

MTB > READ C1 C2
DATA> 3.5    4.0
DATA> END
MTB > NAME C1 '' C2 ''
MTB > PRINT C1-C2
 ROW       C1          C2
   1        3.5         4.0
```

Both upper and lower case letters can be used in a name. When you use a name in a Minitab command, Minitab does not distinguish between upper and lower case letters. However, when the name is printed, upper and lower letters are distinguished, and they

14

follow the initial entry of the name in the NAME command. In the following example you can use 'Height' or 'HEIGHT' after the PRINT command. When C1 is printed, it is labeled Height.

```
MTB > NAME C1 'Height'
MTB > PRINT 'HEIGHT'
```

1.10.4 The PRINT Command

Syntax:
```
PRINT C,...,C
   FORMAT (format specification)

PRINT E, ...,E
```

The PRINT command displays on screen one or more columns, or any mixture of columns, stored constants, and matrices. (To output the results or data on paper, use the WRITE command to save the file; exit Minitab; then print the file, using a print command of your operating system.) If a FORMAT subcommand is used, only columns may be printed. For details of the FORMAT subcommand, see the section on the SET command.

Unlike inputting data, alpha data can be printed with or without using a FORMAT subcommand. Another difference is between the default output format and the default input format. Most computers require the use of the first character in each line for carriage control and line spacing instructions. Therefore, if you were to override the output format with a FORMAT subcommand and if the output is to be directed later on to a printer, then it is best to avoid printing data in the first space of the output line. '1X' at the beginning of the FORMAT statement will produce the desired space.

```
MTB > PRINT C2-C6;
SUBC> FORMAT (1X, 5F3.0).
```

If you mix columns, constants, and matrixes in one PRINT command, they are printed in the order of: matrices, constants, and columns.

To print one column vertically with the row numbers, print the desired column and an empty column.

```
MTB > PRINT 'Final' C20

  ROW   Final   C20
    1      90
    2      89
    3      82
    4      60
    5      78
```

15

```
New Minitab Commands

    SET, READ, END
    NAME, NOTE, #
    LET
    PRINT
    HELP
    STOP
    <Ctrl-Break>
```

EXERCISES

1.   You are to create a data base which will be used for exercises at the end of each remaining chapter. Take a survey of your class, using the following questions:

   a.   How many hours do you study each week?
   b.   How many hours do you work each week?
   c.   How many units are you carrying this semester?
   d.   Marital status: single__married__divorced__other__
   e.   Class level: freshman__sophomore__junior__senior__
   f.   On a scale of 1 to 5, how difficult do you anticipate this class to be?
   g.   What grade do you expect to obtain for this class?
   h.   What is your GPA?
   i.   On a scale of 1 to 10, how satisfied are you with college life?
   j.   On the same scale, indicate your satisfaction with high school life?

2.   After you have collected the data, develop a coding scheme for categorical variables, and determine whether each variable should be numeric or alphabetic.

3.   Using the SET and READ commands, create a Minitab worksheet, and PRINT it. Use various function keys and other facilities to practice editing your data.

CHAPTER TWO
DATA ENTRY AND TABULATION

In This Chapter You Will Learn To:

1. Enter data using the Data Entry Utility
2. Edit and transform columns and rows of data
3. Tabulate sample data

Having gained an overview of Minitab, you are ready to tabulate the sample data on hand. Before discussing the data tabulation, however, this chapter introduces the Data Editor to facilitate data entry, and commands for transforming columns and rows of data.

2.1 DATA ENTRY UTILITY: DATA EDITOR

In the previous chapter you experienced the tedium of correcting data entries with the LET command. This chapter introduces you to the Data Editor, a full screen editor, which facilitates and expedites entry and editing of data.

2.1.1 Data Editor Mode versus Minitab Screen

You can move back and forth between the Data Editor and the Minitab session by simply pressing the ESC key.

Let us see how it works:

1. At the MTB> prompt enter final exam scores into C10 by the SET command as follows:

    ```
    MTB > SET final exam scores into C10
    DATA> 90  89  *  82  60  75  65  70  75
    DATA> END
    MTB > NAME C10 'Final'
    ```

2. Press ESC to shift into the Data Editor mode. The screen is cleared and the spreadsheet (Exhibit 2.1) appears, holding data only in Column 10. At the top of the screen both column names and numbers are displayed. At the bottom of the screen the last column (C10) and row (9) containing data are indicated. However, the row number is hidden by the message: "<F1> for HELP". This message disappears when you press any key.

The little right arrow symbol at the upper left corner shows that the row-wise data entry mode is in effect, which will be explained later.

3.   Now press <u>ESC</u>. The spreadsheet disappears and the original Minitab commands reappear.

Exhibit 2.1

```
                                        Final
    ->      C1  C2  C3  C4  C5  C6  C7  C8  C9  C10  C11  C12  C13  C14  C15
         1                                       90
         2                                       89
         3                                        *
         4                                       82
         5                                       60
         6                                       75
         7                                       65
         8                                       70
         9                                       75
        10
        11
        12
        . . .
        22
                                 Last Column:C10  LAST Row <F1> for HELP
```

2.1.2 Entering New Data

The following is a data set which you will learn to enter into the Minitab worksheet by using the Data Editor. This set contains some errors, which will be corrected by the end of this section.

Exhibit 2.2

Name C1	Sex C2	Age C3	Major C4	GPA C5	Study hours C6	C7	Exam1 C8	Exam2 C9	Final C10
Abbot,John	M	19	1	3.50	20		90	95	90
Baker,Kathy	F	20	2	3.68	25		100	92	89
XXX		*	2	*	22		*	*	*
Cook,Dianne	M	21	3	2.90	18		80	77	82
Hayes,Fred	M	32	1	2.54	10		65	60	60
Lind,Mary	F	41	3	3.10	23		76	82	75
Moe,Tom	M	19	2	2.82	12		62	70	65
O'Connor,Steve	M	25	1	3.02	16		75	*	70
Smiths,Dick	M	21	2	2.73	15		70	65	75

As you enter and edit data within the Data Editor, you will be learning how to move the cursor around to a desired location, to use function keys on the keyboard, and to access items of the Edit Menu. All of these steps are summarized in the Quick Reference Section, Section 2.1.6. As you work through the following exercise, refer to this section any time you feel lost.

1.   Press <u>ESC</u> to shift from the MTB> prompt into the Data Editor.

2.   The cursor is at (Row 1, Column 1). This cell with the cursor blinking in it is called the <u>active cell</u> where datum is entered or edited.

3.   As mentioned earlier, the data entry direction is row-wise, indicated by the right arrow at the upper left corner of the screen. You can change the direction between row-wise and column-wise entry by pressing <u>F3</u>. For now, we will work in the row-wise direction.

      The keys labeled F1,F2,F3...on your keyboard are known as function keys. Refer to the Quick Reference for function keys included in Section 2.1.6.

4.   Type <u>Abbot,John</u> in (Row 1, Col 1). Since the default column width is 4, "Abbot,John" will spread over C1 and C2 to C3. Don't worry about it. When you press <u>ENTER</u>, the width of Column 1 is automatically adjusted to include the full name.

5.   The cursor is now in (Row 1, Col 2). Type <u>M</u>, and this time, press the right arrow key on your keyboard. Both <u>ENTER</u> and <u>right arrow</u> accept the new datum in the cell, and move the cursor to the cell to the right because the data entry direction is row-wise.

6.   Enter the rest of the data for the first row.

7.   Press <u>F6</u> or <u>CTRL-ENTER</u>, which will move the cursor to (Row 2, Col 1). These keys move the cursor to the beginning of the next row if row-wise entry is in effect. If column-wise entry is in effect, they move the cursor to the first cell of the next column. Refer to Section 2.1.6 for cursor movements.

8.   Press <u>F3</u> to change the direction to column-wise. Type <u>Baker,Kathy</u> in (Row 2, Col 1), and press <u>ENTER</u>. Note that the column width is automatically expanded as you enter a longer name than those already in the same column. The cursor now moves to Row 2 and Col 2.

9.   Enter the rest of the data as shown in Exhibit 2.2.

<u>Naming Columns</u>

10.   Press <u>HOME</u> to move the cursor to the first beginning of the worksheet (Row 1, Col 1).

11.   Move the cursor to the column name area by (1) pressing the up arrow key, or (2) pressing <u>F10</u> to display the Edit Menu, and select the "Name" option. When you press F10, the five items of the Edit Menu: Format, GoTo, Help, Name and

Reformat, are listed on the first line of the bottom of the screen. The second line shows the item available for selection at the moment. If it is not the item that you desire, follow the instruction on the third line which is to use the arrow key or to type in the first letter to move to the item of choice.

The column name area is now active. A message "ESCape to exit column name editing; F8 to delete column name" appears at the bottom of the screen.

12. Type <u>Name</u> for C1, and press <u>ENTER</u> to move to C2. Column width is adjusted if you enter a column name which is longer than the existing data in that column. However, column names cannot be longer than eight characters including spaces as you may recall from Chapter 1. Otherwise they will be truncated. Try typing 'study hours' for column 6 as in Exhibit 2.2 and it will be shortened to 'study ho'. Let's rename this column 'work_hrs'.

13. After naming all the columns as in Exhibit 2.2, press <u>ESC</u> or down arrow to return to the data area, as the message at the bottom of the screen instructs you to do.

2.1.3 Editing Data

<u>Editing Characters</u>

1. You can edit data by moving the cursor to a desired location, and reenter a new value. However, more efficient data transformation can be performed by using the Edit mode.

   Move the cursor to (Row 4, Col 2). Press <u>F2</u> to shift into Edit mode. The word "Edit" is displayed at the top and "INSERT" at the bottom of the screen. Press the <u>INS</u> key to shift into Overwrite mode, upon which the "Insert" message at the bottom of the screen disappears. Type <u>F</u> to replace "M" in (Row 4, Col 2), and press <u>ENTER</u>. Note that the "Edit" label has disappeared from the screen.

   To edit another cell, you must first move the cursor to the desired cell or location, and then press <u>F2</u> again.

2. Move the cursor to (Row 7, Col 1), and press <u>F2</u>. Then move the cursor within the cell to "e", type <u>or</u>, and press <u>ENTER</u>. Since INSERT is operative as indicated by the message at the bottom of the screen, "or" is inserted and the name now reads "Moore,Tom". Refer to the Quick Reference for character cursor movements.

3. Move the cursor to (Row 9, Col 1), and press <u>F2</u>. Move the cursor within the cell to "s" before the comma, and press <u>DEL</u> to delete the character. Alternatively, move the cursor to the comma, and press <u>BACKSPACE</u> to delete the character to the left. The name now reads "Smith,Dick".

4.   To jump to a remote cell, the GOTO option is available. Press F10 to display the Editor Menu. Now use the right arrow key and then press ENTER, or simply type G anywhere in the menu to select the Goto option. At the bottom of the screen, a message appears asking you to indicate the column and row numbers of the destination. Type 8 for column, then press ENTER, and 8 for row, then press ENTER.

With the cursor in (Row 8, Col 8), press F2 to shift to Edit mode. Then press INS to change to Overwrite mode, type 80, and press ENTER. Steve O'Connor's score for 'Exam1' is changed from 75 to 80.

Inserting and Deleting Cells

5.   Move the cursor to (Row 8, Col 1). Type Ray,Alex over "O'Connor,Steve", and press F7. Now "Ray,Alex" occupies (Row 8, Col 1), and "Smith,Dick" moves down to (Row 9, Col 1). Note, however, that the data in other columns (C2-C10) have not changed at all.

6.   Let us remove "Ray,Alex" from (Row 8, Col 1). With the cursor in this cell, press F8 to delete the cell. "Smith,Dick" moves up to (Row 8, Col 1).

Inserting and Deleting Rows

7.   You can insert or delete items individually for an entire row, but there is a faster way if you wish to insert or delete a whole row.

Position the cursor anywhere on Row 9, and press Shift-F7. The data in the previous Row 9 is moved down to Row 10, and a new row with missing values is inserted. C1 is blank because it is the alpha column. Other numerical columns have * for missing values (from Chapter 1). Enter values as follows:

C1	C2	C3	C4	C5	C6	C8	C9	C10
Ray,Alex	M	22	1	3.30	20	90	88	85

8.   Move the cursor to Row 3, and press Shift-F8 to delete the third row.

Compressed Mode

9.   Currently Column 7 is empty in the worksheet. To save space on the screen, the display of empty columns can be suppressed by pressing F4. The message "COMPRESS" appears at the bottom of the screen.

10. When you press F4 again, Column 7 reappears on the screen.

2.1.4 Formatting Data

1. To use your own format rather than the default, press F10 to display the Edit Menu, and type F to select the Format option.

2. A format editing area is displayed at the bottom of the screen, containing the formats currently in effect. (Review at this point Chapter 1, Section 1.9.1, the SET Command, if you do not understand the formats displayed at the bottom of the worksheet.) When the default format is used, the message "Auto" is shown below the format. When a column is empty as C7, the message "None" is displayed in place of a format.

3. Currently GPA in C5 is in the default format of F5.2. Move the cursor to the C5 format location by pressing the right arrow key several times. Type F4.1, and press ENTER. The format for C5 now reads "F4.1" and "Fixed". "Fixed" is the opposite of "Auto", the former indicating that the format is defined by the user.

   Since the format is changed to F4.1, the column width for C5 has narrowed, and GPA is allowed only one digit following the decimal point. The Data Editor automatically rounds the two decimal places, which were entered for GPA, to one. For example, Kathy Baker's GPA reads 3.7 instead of 3.68. The screen reads as shown in Exhibit 2.3.

4. Precision of the data is not lost from the worksheet. Type F5.2 as a format for C5, and press ENTER. Previous values are restored. Kathy Baker's GPA is 3.68 again.

Exhibit 2.3

		Name	Sex	Age	Major	GPA	Work_hrs		Exam1	Exam2	Final
->		C1	C2	C3	C4	C5	C6	C7	C8	C9	C10
1	Abbot,John		M	19	1	3.50	20		90	95	90
2	Baker,Kathy		F	20	2	3.68	25		100	92	89
3	Cook,Dianne		F	21	3	2.90	18		80	77	82
4	Hayes,Fred		F	32	1	2.54	10		65	60	60
5	Lind,Mary		F	41	3	3.10	23		76	82	75
6	Moore,Tom		M	19	2	2.82	12		62	70	65
7	O'Connor,Steve	M		25	1	3.02	16		80	*	70
8	Ray,Alex		M	22	1	3.30	20		90	88	85
9	Smith,Dick		M	21	2	2.73	15		70	65	75
10											
11											
. . .											
20											
FMT:		A14	A4	F4.0	F5.0	F5.2	F8.0	None	F5.0	F5.0	F5.0
A/F:		Auto	Auto	Auto	Auto	Fix	Auto	Auto	Auto	Auto	Auto

Reformatting

5. Default formats are automatically adjusted when new data which are wider than the existing field are entered. Thus a 6-digit value "123456" will be entered by a F7.0 format when a smaller field is in effect. However, when a large value is removed from a column, the column format is not adjusted automatically. Reformatting is necessary to restore the default format.

6. Press ESC as instructed at the bottom of the screen to exit format editing mode. Enter 123 in Row 1 Col 11; then type 123456 in Row 2, Col 11, and press ENTER. Press F10 to display the Edit Menu, and select the Format option. C11 shows a format F7.0 to include the larger value plus a space in between the columns.

7. With the cursor on Row 2, Col 11, press F8 to delete the value "123456".

8. Press F10 to display the Edit Menu, and select the Format option. The format for C11 is still F7.0, even though a format of F4.0 is sufficient to contain the first value of 123.

9. Exit format editing mode by pressing ESC, then press F10 to display the Edit Menu again, and type R to select the Reformat option. Reformatting is carried out for the column at which the cursor is located. So make sure that your cursor is on C11.

10. Press F10 to return to the Edit Menu, and select the Format option to confirm that reformatting has taken place. The format for C11 is now F4.0. Delete the value 123 in Col 11 before going on to the next section.

Alpha Data

11. If you enter characters in a column, that column is defined automatically as an alpha column. However, if you wish to treat numbers such as Social Security Numbers as alpha data, you must define the column as alpha with a fixed format before entering the data.

12. Press F10 for the Edit Menu, and select the Format option. Change the format for C11 to A9, and press ENTER.

13. Enter Social Security Numbers in C11 for everyone.

2.1.5 Returning to the Command Line

1. Press ESC to return to the MTB> prompt.

2.   Fill in C7 with another variable, 'Reason', by using the SET command.

```
MTB > SET C7
DATA> 3 3 1 1 2 1 1 3 4
DATA> END
MTB > NAME C7 'Reason'
```

'Reason' , referring to the reason for taking this course, is coded as follows:

1.   It is required for my major.
2.   I have heard great things about the instructor.
3.   I am interested in the subject matter of the course.
4.   I could not get into any other class at this hour
5.   Other reasons

2.1.6 Quick Reference

<u>Cursor Move in Enter Mode</u>

ENTER	Accept new data, and move one cell in the direction indicated, row-wise or column-wise
Arrow keys	The same as ENTER
BACKSPACE	Delete the character to the left of the cursor
CTRL-ENTER or F6	Accept new data, and move the cursor row-wise or column-wise
PageUp, PageDn	Move up or down one page of the worksheet
CTRL-left arrow	Move left one page
CTRL-right arrow	Move right one page
HOME	Move to the upper left corner of the screen
END	Move to the lower right corner of the screen
CTRL-HOME	Move to the upper left corner of the worksheet
CTRL-END	Move to the lower right corner of the worksheet

<u>Cursor Move in Edit Mode</u>

Left, right arrow	Move the cursor one character left or right
HOME	Move to the beginning of the cell
END	Move to the end of the cell
BACKSPACE	Delete the character to the left of the cursor
DELETE	Delete the character at the cursor
INSERT	Switch between inserting and overwriting modes

Function Keys

F1	Help screen
F2	Edit mode to edit characters in a cell
F3	Row-wise vs column-wise entry direction
F4	Compressed mode, suppressing the display of empty columns
F6	Same as CTRL-ENTER
F7	Insert a data item
SHIFT-F7	Insert a row
F8	Delete a data item
SHIFT-F8	Delete a row
F10	Editor Menu

Editor Menu Options

Format	Format a column
GoTo	Go to a user-specified column and row
Help	Give information on the Data Editor
Name	Name columns
Reformat	Reformat a column

2.2 SAVING AND RETRIEVING THE DATA

2.2.1 Saving the Worksheet into a File : The SAVE Command

1. After spending so much time in creating a worksheet, you may not want to lose it all when you exit Minitab. There is a command, SAVE, which saves the data in a worksheet format, placing it into a file.

2. Let us save the current worksheet in the file, CLASS.MTW in drive A. The filename can be up to eight characters long, containing letters and/or numbers. The extension MTW is automatically attached. If a filename is not entered, the worksheet is saved in a default file named, MINITAB.MTW.

```
MTB>   SAVE 'A:CLASS'
```

The computer responds with "Worksheet saved into file: A:CLASS.MTW".

2.2.2 Retrieving the Worksheet: The RETRIEVE Command

1. Now that the worksheet containing the class data are safely saved, we can reuse the columns for some other purposes.

SET clears away the existing values from C1, and enters the new values, 10, 20, and 30. The NAME command erases the previous column name for C1. On the other hand, C2 retains the former column name, 'Sex', and its values, as shown below.

```
MTB > SET C1
DATA> 10 20 30
DATA> END
MTB > NAME C1 ''
MTB > PRINT C1-C2
  ROW    C1    Sex
    1     10      M
    2     20      F
    3     30      F
    4             F
    5      .  .  .
```

Let us retrieve the worksheet containing the class data by using the RETRIEVE command. The filename with its drive and path is specified after the command RETRIEVE and are enclosed in parentheses. The filename extension MTW is not necessary.

```
MTB > RETRIEVE 'A:CLASS'
```

2.   When RETRIEVE is executed, all the current data in the worksheet are replaced by the retrieved worksheet.

2.2.3 Displaying the Contents of the Worksheet: <u>The INFORMATION Command</u>

1.   After the CLASS.MTW is retrieved, its contents are not displayed unless you ask the computer to do so. The INFORMATION command summarizes the contents of the current worksheet. It lists the column number, column name (if any), number of rows, and number of missing values (if any). The "A" by the column number indicates that the column contains alpha data.

```
MTB>   INFO

        COLUMN    NAME        COUNT      MISSING
  A       C1      Name          9
  A       C2      Sex           9
          C3      Age           9
          C4      Major         9
          C5      GPA           9
          C6      Work_hrs      9
          C7      Reason        9
          C8      Exam1         9
          C9      Exam2         9          1
          C10     Final         9
```

2. If you specify columns after the command, information concerning the designated columns alone will be displayed.

```
MTB > INFORMATION on C3 C5 C7
```

2.2.4 Writing Data to a File or to the Screen: The WRITE Command

1. The worksheet file, CLASS.MTW, is stored in a special Minitab code which cannot be read by non-Minitab software. In contrast the WRITE command copies the data in the current worksheet into a data file in ASCII, which is readable by other software. To restore a file which is copied with the WRITE command, you must use the READ command.

   Although the WRITE command can stand alone without the subcommand FORMAT, it is advisable to use this subcommand anyway. Then you can READ the file with the same format without trying to figure out the default format used by Minitab.

2. Let us save the current worksheet in an ASCII data file:

```
MTB>  WRITE 'A:CLASS' C1-C10;
SUBC> FORMAT (A14,1X,A1,1X,F2.0,1X,F1.0,1X,F4.2,1X,F2.0,1X, &
CONT> F1.0,1X,F3.0,2(1X,F2.0)).
```

   This saves the data from columns C1 through C10 of the current worksheet in the file, CLASS.DAT in drive A in the specified format which can then be READ again in the same format. The extension DAT is automatically attached. If no filename is specified, WRITE displays data on the screen.

2.3 COLUMN AND ROW MANIPULATION COMMANDS

You have found how easy it is to edit a worksheet by the Data Editor. However, you should also learn commands which perform various column and row manipulations.

2.3.1 Inserting Rows: The INSERT Command

1. Let's add three students to the bottom of the class file.

```
MTB > INSERT C1-C10;
SUBC> FORMAT(A14,1X,A1,1X,F2.0,1X,F1.0,1X,F4.2,1X,F2.0, &
CONT> 1X,F1.0,3(1X,F3.0)).
DATA> Conrad,Jane    F 20 4 2.95  8 4  88  75  66
DATA> Hardy,Jill     F 25 2 2.88 20 4  67  73  65
DATA> Freund,May     F 20 2 2.95  2 1  65  77  75
DATA> END
```

27

2. The FORMAT subcommand is necessary because there are alpha data to be inserted. In the Data Editor the default format shows (A4) for C2, but on the command prompt level, it is necessary to specify the format as (A1) since the data in C2 are only one space wide.

3. Press <u>ESC</u> to shift into the Data Editor screen. You will see the above three students added at the bottom.

4. Press <u>ESC</u> again to return to the MTB> prompt, and save the updated worksheet.

```
MTB > SAVE 'A:CLASS'
```

5. To insert data in certain columns between rows, you need to specify both the columns and rows. Let us insert "ABC,DEF" between rows 7 and 8.

```
MTB > INSERT between rows 7 and 8, in C1;
MTB > FORMAT(A14).
DATA> ABC,DEF
DATA> END
```

ROW	C1	C2	C3	C4	C5	C6	C7	C8	C9	C10
1	Abbot,John	M	19	1	3.50	20	3	90	95	90
2	Baker,Kathy	F	20	2	3.68	25	3	100	92	89
3	Cook,Dianne	F	21	3	2.90	18	1	80	77	82
4	Hayes,Fred	F	32	1	2.54	10	1	65	60	60
5	Lind,Mary	F	41	3	3.10	23	2	76	82	75
6	Moore,Tom	M	19	2	2.82	12	1	62	70	65
7	O'Connor,Steve	M	25	1	3.02	16	1	80	*	70
8	ABC,DEF	M	22	1	3.30	20	3	90	88	85
9	Ray,Alex	M	21	2	2.73	15	4	70	65	75
10	Smith,Dick	F	20	4	2.95	8	4	88	75	66
11	Conrad,Jane	F	25	2	2.88	20	4	67	73	65
12	Hardy,Jill	F	20	2	2.95	2	1	65	77	75
13	Freund,May									

When an item is inserted in C1 between rows 7 and 8, items below that in C1 are pushed down one row, while the data in C2-C10 remain the same.

2.3.2 Deleting Rows: <u>The DELETE Command</u>

1. Let's remove the odd name we inserted above.

```
MTB > DELETE row 8   in C1
```

The name "ABC,DEF" is deleted from (Row 8, Col 1), and the names below in C1 are moved up one row.

28

2. To delete more than one row in more than one column, specify the rows and columns after the command. The command below will delete the rows containing information for Fred Hayes and Mary Lind in all columns.

```
MTB > DELETE rows 4 5  in columns C1-C10
```

2.3.3 Erasing Columns and Constants: The ERASE Command

1. To remove columns or constants from the worksheet, you must use the ERASE command.

```
MTB > ERASE C8-C9
```

2. When the worksheet is displayed by the PRINT command, columns 8 and 9 are empty.

2.3.4 Initializing the Worksheet: The RESTART Command

1. While practicing various row and column manipulation commands, we have botched the original class data worksheet. To clear the current worksheet there is a command, RESTART, which erases all the columns and rows.

```
MTB > RESTART
MTB > PRINT C1-C10
```

After executing the RESTART command, the succeeding PRINT command generates a message that the worksheet is currently empty.

2. You can begin a new Minitab session at this point.

2.4 TABULATING DATA

2.4.1 Sorting the Data: The SORT Command

1. Let's retrieve the class data into a cleared worksheet by the RETRIEVE command. Since three students were added at the bottom of the file, we need to sort the worksheet by name so that the students are arranged in an alphabetic order.

2. The SORT command sorts specified columns (C1-C10) by rows, according to the values in the first column listed (C1), and in ascending order, with the lowest value listed first and the highest value last. Sorting alphabetically in ascending order means that A's will come before B's and so forth. The rearranged values are stored in a

second series of columns (C1-C10), which can be and are in this example the same as the original columns. Thus the first specification of columns after SORT refers to the columns to be sorted, while the second specification designates a new storage location for the sorted values.

```
MTB > RETRIEVE 'A:CLASS'
MTB > #  Sort by name in alphabetical order
MTB > SORT C1-C10 C1-C10
MTB > PRINT C1-C5
```

ROW	Name	Sex	Age	Major	GPA
1	Abbot,John	M	19	1	3.50
2	Baker,Kathy	F	20	2	3.68
3	Conrad,Jane	F	20	4	2.95
4	Cook,Dianne	F	21	3	2.90
5	Freund,May	F	20	2	2.95
6	Hardy,Jill	F	25	2	2.88
7	Hayes,Fred	F	32	1	2.54
8	Lind,Mary	F	41	3	3.10
9	Moore,Tom	M	19	2	2.82
10	O'Connor,Steve	M	25	1	3.02
11	Ray,Alex	M	22	1	3.30
12	Smith,Dick	M	21	2	2.73

3. Next, sort students according to sex first, then within males and females sort according to the final exam scores.

When the BY subcommand is used, data are sorted first, according to the first column, ('Sex'), then, by the second column ('Final') specified after BY.

```
MTB > #  Sort first by Sex, then by Final exam scores
MTB > SORT C1-C10 C1-C10;
SUBC> BY 'Sex' 'Final'.
MTB > PRINT 'Sex' 'N`me' 'Final'
```

ROW	Sex	Name	Final
1	F	Hayes,Fred	60
2	F	Hardy,Jill	65
3	F	Conrad,Jane	66
4	F	Freund,May	75
5	F	Lind, Mary	75
6	F	Cook,Dianne	82
7	F	Baker,Kathy	89
8	M	Moore,Tom	65
9	M	O'Connor,Steve	70
10	M	Smith,Dick	75
11	M	Ray,Alex	85
12	M	Abbot,John	90

4. Next sort students according to the final exam score from the highest to the lowest, and store the results in columns 14 and 15. Print the exam scores followed by students' names. The subcommand DESCENDING is needed to effect this sorting. Data are sorted according to the column named after DESCENDING.

```
MTB > SORT 'Final' 'Name' C14-C15;
MTB > DESCENDING 'Final'.
MTB > NAME C14 'Position' C15 'Student'
MTB > PRINT C14-C15
```

ROW	Position	Student
1	90	Abbot,John
2	89	Baker,Kathy
3	85	Ray,Alex
4	82	Cook,Dianne
5	75	Lind,Mary
6	75	Smith,Dick
7	75	Freund,May
8	70	O'Connor,Steve
9	66	Conrad,Jane
10	65	Moore,Tom
11	65	Hardy,Jill
12	60	Hayes,Fred

2.4.2 Ranking Scores: The RANK Command

1. The RANK command assigns rank scores to values in a specified column. Let's rank GPAs, then sort students by the ranks of GPA.

```
MTB > RANK data in 'GPA' put ranks in C11
MTB > NAME C11 'Rank'
MTB > SORT 'Rank' 'Name' 'Rank' 'Name'
MTB > PRINT 'Rank' 'Name'
```

ROW	Rank	Name
1	1.0	Hayes,Fred
2	2.0	Smith,Dick
3	3.0	Moore,Tom
4	4.0	Hardy,Jill
5	5.0	Cook,Dianne
6	6.5	Conrad,Jane
7	6.5	Freund,May
8	8.0	O'Connor,Steve
9	9.0	Lind,Mary
10	10.0	Ray,Alex
11	11.0	Abbot,John
12	12.0	Baker,Kathy

A rank of 1 is assigned to the lowest GPA, 2 to the second lowest value and so on. Note that Conrad,Jane and Freund,May have the same GPA of 2.95. They are therefore assigned the tied rank of 6.5, which is the average of the ranks 6 and 7, which they would have occupied had they not been tied.

2.4.3 Frequency Distribution: <u>The TABLE Command</u>

1. Sorting accompanied by printing is a good way of tabulating selected information according to a desired classification variable. However, the next step in tabulation is to generate a frequency distribution table.

    The TABLE command followed by column specification displays a count of the observations for each value in the column. To obtain relative frequencies as well as counts, the subcommands COUNTS and TOTPERCENTS are required.

    ```
    MTB > NOTE  Calculate the Relative Frequencies of Reasons
    MTB > TABLE for 'Reason';
    SUBC> COUNTS;
    SUBC> TOTPERCENTS.

     ROWS: Reason

            COUNT % OF TBL

      1         5      41.67
      2         1       8.33
      3         3      25.00
      4         3      25.00
     ALL       12     100.00
    ```

2. A twoway table, i.e., crosstabulation of two variables, is produced by listing two columns after the command keyword, TABLE. The first column is the row variable, the second the column variable. A cross tabulation of 'Reason' by 'Major' is shown below.

    ```
    MTB > TABLE 'Reason' by 'Major';
    SUBC> COUNTS;
    SUBC> TOTPERCENTS.
    ```

```
ROWS: Reason        COLUMNS: Major

                  1         2         3         4        ALL

        1         2         2         1         0          5
               16.67     16.67      8.33        --       41.67

        2         0         0         1         0          1
                  --        --      8.33        --        8.33

        3         2         1         0         0          3
               16.67      8.33        --        --       25.00

        4         0         2         0         1          3
                  --     16.67        --      8.33       25.00

      ALL         4         5         2         1         12
               33.33     41.67     16.67      8.33      100.00
      CELL CONTENTS  --
                        COUNT
                        % OF TBL
```

2.4.4 Cumulative Frequencies and Percentages: <u>The TALLY Command</u>

1. Another command that produces frequency distribution is the TALLY command.
    TALLY with COUNTS and PERCENTS subcommands produces the same result as
    TABLE described above.

    The main difference is that TALLY can tabulate more than one column at a time,
    outputting multiple columns across the output screen. Another advantage of TALLY
    is that it provides cumulative counts, CUMCNTS, and percentages, CUMPCTS.

2. When all four subcommands are requested (COUNTS, PERCENTS, CUMCNTS,
    and CUMPCTS), you may simply substitute the subcommand ALL.

```
MTB > TALLY 'Age';
SUBC>    COUNTS;              # Frequency counts
SUBC>    CUMCNTS;             # Cumulative counts
SUBC>    PERCENTS;            # Relative percentages
SUBC>    CUMPCTS.             # Cumulative relative percentages

         Age   COUNT   CUMCNT   PERCENT  CUMPCTS
     19    2       2    16.67    16.67
     20    3       5    25.00    41.67
     21    2       7    16.67    58.33
     22    1       8     8.33    66.67
     25    2      10    16.67    83.33
     32    1      11     8.33    91.67
     41    1      12     8.33   100.00
      N=  12
```

2.4.5 Frequencies for Grouped Data: The CODE Command

1. Sometimes you may wish to simplify the data by collapsing them into fewer categories. For example, scores are normally converted into letter grades of A for 91-100, B for 81-90, and so on. The CODE command performs this type of recoding, thus creating a new variable with different and possibly fewer codes.

2. After the command keyword, CODE, specify a value or a range of values to be changed in parentheses, which is followed by the new code. After all the old and new values are designated, then indicate the input column(s) to be recoded and an equal number of output column(s) to store the new variable(s) created.

   In the example below, scores of the final exam are recoded so that values in the range 91-100 are coded as 4, those between 81 and 90 are coded as 3, and so on. The input column is 'Final' and the output column is C12.

```
MTB > CODE (91:100)4 (81:90)3 (71:80)2 (61:70)1 (0:60)0    &
CONT> 'Final',put in C12
MTB > NAME C12 'Grade'
MTB > TALLY C12;
SUBC>    ALL:
```

Grade	COUNT	CUMCNT	PERCENT	CUMPCT
0	1	1	8.33	8.33
1	4	5	33.33	41.67
2	3	8	25.00	66.67
3	4	12	33.33	100.00
N=	12			

3. CODE can be used only to transform numerical variables, not alpha data. The new codes to be generated must also be either numerical or missing value. In the following example the missing value code '-99' is replaced by the missing value symbol '*'.

```
MTB > CODE (-99) '*' C20 C21
```

2.5 MORE ON COMMANDS

2.5.1 The SAVE Command

Syntax:
```
SAVE [in 'filename'] a copy of the worksheet
    PORTABLE
```

The SAVE command saves all the columns, data, column names, and constants from the current worksheet into a file with the default extension, MTW. The resulting file can be retrieved by the RETRIEVE command. SAVE handles alpha columns as it does numeric columns.

The PORTABLE subcommand saves the worksheet in a format that can be transferred to a computer of a different type (IBM/PC, VAX, etc.). The default filename extension for the portable file is MTP.

2.5.2 The WRITE Command

Syntax:

```
WRITE [to 'filename'] the data in columns C...C
   FORMAT (format statement)
```

The WRITE command writes data to a file or to the screen, although it is normally used to create a data file. The command keyword, WRITE, is followed by the specification of the filename and the columns where the data are found.

The optional FORMAT subcommand followed by the format statement in parentheses specifies where and how the data should be output.

The file created by the WRITE command may be used with the READ, SET or INSERT commands as well as being read by software other than Minitab. The default file extension for a file created by this command is DAT.

WRITE command can be used for alpha data with or without a FORMAT subcommand.

There is no carriage control in the FORMAT for WRITE.

```
MTB > WRITE 'B:LETTER' C1-C3;
SUBC> FORMAT(A3,2(3X,2F1.0)).
MTB > TYPE B:LETTER

  ABC    1   2
  XYZ    3   4
```

Missing data are output as a field with an * in the rightmost space.

35

2.5.3 The RETRIEVE Command

Syntax:

```
RETRIEVE    ['filename']
PORTABLE
```

The RETRIEVE command only restores a MTW file created by the SAVE command. If you try to retrieve an ASCII file, you will receive an error message that the file is not in a Minitab format.

Note also that this command clears away all the data in the current worksheet before retrieving the specified file. Be sure to save the current worksheet before retrieving another.

The filename must be enclosed in single quotation marks, and accompanied by the drive name and the path (if necessary).

The PORTABLE subcommand allows retrieval of a file from other computers such as the mainframe. It retrieves a worksheet SAVEed with the PORTABLE subcommand on a different computer type.

2.5.4 The INSERT Command

Syntax:

```
INSERT   ['filename'] between K and K of C,...,C
INSERT   ['filename'] at the end of C,...,C

FORMAT   (format specification)
NOBS = K
```

The INSERT command inserts additional rows of data into the existing worksheet. Additional rows may come from a file specified or from the keyboard at the DATA prompt, following the INSERT command.

To put the data at the top of the columns, specify the rows as:

INSERT ['filename'] between rows 0 and 1 of C,...,C

To add rows at the end of the columns, omit the row specification.

Alpha data are inserted into columns with a FORMAT subcommand in the same way as in the SET command.

The FORMAT and NOBS subcommands work in the same way as the corresponding subcommands for READ (with more than one column) and SET (with one column).

2.5.5 The DELETE Command

Syntax: | DELETE   rows K...K in columns C...C |

The DELETE command deletes specified rows from columns in the current worksheet, and adjusts the space allocation to fill in gaps.

2.5.6 The SORT Command

Syntax: | SORT values in C [carry along corresp. rows of C...C]
put into C [pur corresp. rows into C...C]

By C
DESCENDING C |

The SORT command sorts one or more columns in ascending or descending order. The default, without the BY subcommand, is to sort by the first column in ascending order. For multiple-column sorting the BY subcommand is necessary. Data will be sorted by the first, by the first column specified after "BY", then, by the second column specified, and so on. SORT handles any combination of alpha or numeric columns.

Missing values, *, in numeric columns are sorted last, while missing values in alpha columns, blanks, are sorted first.

2.5.7 The TABLE Command

Syntax: | TABLE data classified by C,...,C
COUNTS - frequency counts for each entry
TOTPERCENTS - percentages of the total table |

The TABLE command produces oneway, twoway, and multiway tables. If one column is specified after TABLE, oneway tabulation is produced. Specification of two columns results in a cross tabulation of two variables. The values in the column must be integer values.

If TABLE has no subcommands, each cell contains just the count of the observations in the cell. To obtain the relative frequencies the subcommand TOTPERCENTS is needed. There are many more subcommands which produce various statistics, the discussion of which is postponed till later. Summary statistics will be presented in Chapter 3, and the chisquare test in Chapter 8.

The FREQUENCIES subcommand is used when the raw data are already collapsed into frequencies by certain categories. For example, suppose we have data for another class concerning the reasons students take the class. Answers are already tabulated, but you would like to compute percentages.

```
MTB>    NOTE C21: Reasons for Students' Taking a Given Course
MTB>    NOTE C22: Numbers of Persons in Statistics Course
MTB>    READ C21-C22
MTB>    1    20
MTB>    2     9
MTB>    3     5
MTB>    4     4
MTB>    5     8
MTB>    END
MTB>    NAME C21 'Reason' C22 'Stat'
```

Given the above frequency distribution, you can compute percentages as follows:

```
MTB>    TABLE FOR 'Reason';
SUBC>   FREQUENCIES IN 'Stat';
SUBC>   COUNTS;
SUBC>   TOTPERCENTS.
   ROWS: Reason
                 COUNT    % OF TBL

       1          20        43.48
       2           9        19.57
       3           5        10.87
       4           4         8.70
       5           8        17.39
     ALL          46       100.00
```

2.5.8 The TALLY Command

```
Syntax:  ┌─────────────────────────────────────────────────┐
         │ TALLY data in C,...,C                           │
         │    COUNTS - frequency counts                    │
         │    PERCENTS - relative percentages              │
         │    CUMCNTS - cumulative counts                  │
         │    CUMPCTS - cumulative relative percentages    │
         │    ALL -     four statistics above              │
         └─────────────────────────────────────────────────┘
```

The TALLY command with the subcommands COUNTS and PERCENTS produces the same result as the TABLE command. The main difference is that TALLY can tabulate and display multiple columns across the output screen. Another advantage of TALLY is that it provides cumulative counts and cumulative percentages.

```
NEW MINITAB COMMANDS

   Data Editor Commands
   SAVE, WRITE, RETRIEVE, INFORMATION
   INSERT, DELETE, ERASE, RESTART
   SORT, RANK, TABLE, TALLY, CODE
```

EXERCISES

1. Using the Data Editor, complete the data entry for the entire class which you started in Chapter 1. Save it in a worksheet format and also in an ASCII file.

2. Practice data transformation using the Data Editor and commands at the prompt.

3. Classify students into two groups according to the number of hours of study.

4. Sort the class by 'St_type' and GPA. Print 'St_type', GPA, and student's name.

CHAPTER THREE
DESCRIPTIVE STATISTICS

```
In This Chapter You Will Learn To:

    1.  Use commands and functions for descriptive statistics
            Arithmetic operation
            Descriptive statistics of rows and columns
    2.  Manipulate rows and columns
            COPY, STACK, UNSTACK, CONVERT, CONCATENATE
```

Having tabulated the data in the previous chapter, we know the shape of the distribution. The next logical step is to find out the average score, the standard deviation, and other summary measures to describe the characteristics of the data. This chapter introduces various Minitab functions and commands which produce descriptive statistics. In addition, commands to manipulate rows and columns are presented as they are useful in computing statistics for selected portions of data.

3.1 ARITHMETIC OPERATIONS

3.1.1 Value Assignment: <u>The LET Command</u>

Syntax:
```
LET   E = expression
```

The LET command can assign any value or a mathematical expression to a new or existing column, or a stored constant. Arithmetic operations available are shown below in the order of hierarchy:

1.	**	exponentiation			
2.	*	multiplication	or	/	division
3.	+	addition	or	-	subtraction

Basic arithmetic rules apply here such that the expression within parentheses is performed first, followed by exponentiation, then by multiplication and division, and finally by addition and subtraction.

You can also use comparisons and logical operators in conjunction with LET. For comparisons the following symbols are available:

EQ	=	LT	<	LE	<=
GT	>	GE	>=	NE	˜=

Logical operators are:

AND	&	OR	¦	NOT	˜

The value of 1 is assigned if the outcome of the comparison or logical operation is true. Otherwise, the value is set to 0. Missing values can also be used in these operations.

No extra text may be added on a LET command except after the # symbol. The LET command, used with arithmetic functions, is described in Section 3.1.3.

1.   The following example computes the average score of the two mid-term exams for each student. Note that when a missing value is found in a column, computation is not performed. Only * appears in the output column.

```
MTB > RETRIEVE 'A:CLASS'
MTB > LET C11=(C8+C9)/2
MTB > NAME C11 'Average'
MTB > PRINT C8 C9 C11
```

ROW	Exam1	Exam2	Average
1	90	95	92.5
2	100	92	96.0
3	80	77	78.5
4	65	60	62.5
5	76	82	79.0
6	62	70	66.0
7	80	*	*
8	90	88	89.0
9	70	65	67.5
10	88	75	81.5
11	67	73	70.0
12	65	77	71.0

2.   In the following example weights are given to various exams to compute the final grade. For example, the weight of .25 is given to both 'Exam1' and 'Exam2' and 'Final' is assigned a weight of .5.

```
MTB > LET C11=C8*.25 + C9*.25 + C10*.5
MTB > NAME C11 'Total'
MTB > PRINT C8 C9 C10 C11
```

41

```
ROW   Exam1   Exam2   Final    Total

 1       90      95      90     91.25
 2      100      92      89     92.50
 3       80      77      82     80.25
 .  .  .
 7       80       *      70        *
 .  .  .
12       65      77      75     73.00
```

3.1.2 Simple Arithmetic Operations

Syntax:

```
ADD        E to E to E...,      put into E
SUBTRACT   E from E,            put into E
MULTIPLY   E by E ... BY E,     put into E
DIVIDE     E by E,              put into E
RAISE      E to the power E,    put into E
```

The above commands do arithmetic on columns and constants. For each E in the command description you can use either a column (such as C1), a stored constant (such as K1), or a number (such as 5).

If any row of an input column contains a missing value, *, the corresponding row of the outcome column is defined as *.

1. In the following example, the scores in C8 (Exam1) and C9 (Exam2) are summed, and the result is stored in column 12.

```
MTB > ADD 'Exam1' to 'Exam2', put in C12
MTB > NAME C12 'Sum_Mid'
MTB > PRINT 'Exam1' 'Exam2' 'Sum_Mid'

ROW   Exam1   Exam2   Sum_Mid

 1       90      95       185
 2      100      92       192
 3       80      77       157
 .  .  .
 7       80       *         *
 .  .  .
12       65      77       142
```

2. Now, let's assign 100 to a stored constant, K1. Then subtract the final exam scores from K1 to see how far removed each exam score is from the full point of 100.

```
MTB > LET K1=100
MTB > SUBTRACT C10 from K1, put in C13
MTB > NAME C13 'Diff'
MTB > PRINT K1 'Final' 'Diff'
K1          100.000

   ROW   Final    Diff

     1      90      10
     2      89      11
     3      82      18
   . . .
    12      75      25
```

3.   In the following example we are performing the same task as we have done earlier: giving weights to different exams and summing the scores. Let's see if we obtain the same results.

```
MTB > MULTIPLY 'Exam1' BY .25 put in C14
MTB > MULTIPLY 'Exam2' by .25 put in C15
MTB > MULTIPLY 'Final' by .50 put in C16
MTB > ADD C14 to C15 to C16 put in C17
MTB > NAME C17 'Total2'
MTB > PRINT 'Exam1' 'Exam2' 'Final' 'Total2'

   ROW   Exam1   Exam2   Final   Total2

     1      90      95      90    91.25
     2     100      92      89    92.50
     3      80      77      82    80.25
            . . .
     7      80       *      70        *
            . . .
    12      65      77      75    73.00
```

4.   In the following example divide the number of working hours per week by 5 so that average daily working hours are displayed together with students' ages.

```
MTB > DIVIDE 'Work_hrs' by 5, put into C18
MTB > NAME C18 'Day_wk'
MTB > PRINT 'Age' 'Day_wk'

   ROW   Age   Day_wk

     1    19      4.0
     2    20      5.0
     3    21      3.6
   . . .
    12    20      0.4
```

43

3.1.3 Arithmetic Functions

1. Minitab has a built-in logical and mathematical <u>function</u>, which can be used together with the LET command.

    We have computed the sum of weighted exam scores (Total2), which has fractions. Let's round it, using the ROUND function. After the function name, ROUND, specify the argument, which is a column or a stored constant in parentheses. The function with the argument is then assigned to a column or a stored constant by LET.

    ```
    MTB > LET C19=ROUND('Total2')
    MTB > NAME C19 'Grade'
    MTB > PRINT 'Name' 'Grade'

       ROW            Name   Grade

         1     Abbot,John        91
         2     Baker,Kathy       93
         3     Cook,Dianne       80
               . . .
         7     O'Connor,Steve     *
               . . .
        12     Freund,May        73
    ```

    Some of the frequently used functions with the LET command are shown below:

ABSOLUTE	Absolute value of the argument
SQRT	Square root of the argument
LOGTEN	Log to the base 10 of the argument
LOG	Natural log
EXPO	e to the power of the argument
ANTILOG	10 to the power of the argument
ROUND	Rounding the argument to the nearest integer
SIN	Sine of the argument in radians
COS	Cosine of the argument in radians
TAN	Tangent of the argument in radians
ASIN,ACOS,ATAN	the arc sine, arc cosine, arc tangent
SIGNS	+1 for a positive argument
	-1 for a negative argument
	0 for a zero value
NSCORES	Normal scores for the data

3.2 SUMMARIZING DATA IN COLUMNS OR ROWS

3.2.1 Column-Wise Functions

Syntax:
```
COMMAND value of E, put into E
```

The following are commands which compute the designated functions without the use of LET. Note that as commands, they have arguments which are not enclosed in parentheses. The arguments which follow such functions can be a column, a number, or a stored constant.

ABSOLUTE	SQRT	LOGE	LOGTEN	EXPONENTIATE
ANTILOG	ROUND	SIN	COS	TAN
ASIN	ACOS	ATAN	SIGNS	

PARSUMS   (partial sums)
PARPRODUCTS (partial products)
NSCORES   (normal scores)

1. There are two mid-term exam scores in the class data. How large is the absolute difference between these two scores?

```
MTB > LET C21='Exam1' - 'Exam2'
MTB > ABSOLUTE C21 in C22
MTB > PRINT 'Exam1' 'Exam2' C21 C22
```

ROW	Exam1	Exam2	C21	C22
1	90	95	-5	5
2	100	92	8	8
3	80	77	3	3
. . .				
7	80	*	*	*
. . .				
12	65	77	-12	12

3.2.2 Column-Wise Statistics

To describe the characteristics of a column, Minitab offers the following useful column-wise statistics.

Syntax:
```
COMMAND in C [put in K]
```

45

The syntax is the same for all of the following column-wise statistics commands. It begins with the command keyword or name, followed by a column specification. If you desire to output the result as a certain stored constant, designate the location of the stored constant, which as you may recall is referred to as K.

COUNT	Number of missing and nonmissing values
N	Number of nonmissing values
NMISS	Number of missing values
SUM	Sum of nonmissing values
MEAN	Mean of the nonmissing values
STDEV	Standard deviation of the nonmissing values
MEDIAN	Median of the nonmissing values
MINIMUM	Smallest nonmissing value
MAXIMUM	Largest nonmissing value
SSQ	Uncorrected sum of the squares of the nonmissing values

1. Let's compute the means of age and GPA. The result appears on the screen immediately upon execution. However, you can assign the mean to a stored constant(K2) to be used later for some other purpose.

```
MTB > MEAN of 'Age'
   MEAN    =      23.750
MTB > MEAN OF 'GPA' store in K1
   MEAN    =      3.0308
MTB > PRINT K1
K1       3.03083
```

2. The average age can also be computed using the MEAN and SUM as arithmetic functions with LET. In the example below, K2 contains the mean computed by using SUM and N as functions, while K3 contains the mean produced by the MEAN function. The two methods should produce the same value, 23.750.

```
MTB > LET K2=SUM('Age')/N('Age')
MTB > LET K3=MEAN('Age')
MTB > PRINT K2 K3
K5       23.7500
K6       23.7500
```

3. As an exercise, compute the standard deviation of the final exam scores by using column-wise statistics. Then compute the same by following the equation of the standard deviation, utilizing all the functions and arithmetic operations you have learned so far.

standard deviation = sqrt[{Σ (X_i- $\bar{X}$)**2}/(N-1)]

46

```
MTB > STDEV of 'Final'
    ST.DEV. =         10.001

MTB > MEAN OF 'Final' K4
    MEAN      =       74.750

MTB > SUBTRACT K4 from 'Final' C11
MTB > RAISE C11 to power 2 C12
MTB > SUM of C12 K5
    SUM       =       1100.3
MTB > LET K6=K5/(N('Final') - 1)
MTB > LET K7=SQRT(K6)
MTB > PRINT K7
K4          10.0011
```

3.2.3 Row-Wise Statistics

Minitab provides descriptive statistics for data in rows as well. For example, you may wish to compute the average of two exams for each person. Exam scores for each person are stored in a row as we have seen so far.

Statistics available for row-wise data are the same as those available for column-wise data. The difference is that an additional "R" is prefixed to column-wise statistics commands. Thus RCOUNT, RN, RNMISS, RSUM, RMEAN, RSTDEV, RMEDIAN, RMINIMUM, RMAXIMUM, and RSSQ are commands which operate on rows. Another difference pertains to the storage of results. Since there is an output for each row, it is necessary to indicate the column where the result should be stored.

1.   Compute the mean of two mid-term exams, and store the result as 'Average'.

```
MTB > RMEAN of 'Exam1' 'Exam2' store in C25
MTB > NAME C25 'Average'
MTB > PRINT 'Exam1' 'Exam2' 'Average'

ROW   Exam1  Exam2  Average

    1      90     95     92.5
    2     100     92     96.0
    3      80     77     78.5
.  .  .
    6      62     70     66.0
    7      80      *     80.0
.  .  .
   12      65     77     71.0
```

3.3 SUMMARIZING DATA IN CELLS

The TABLE command introduced in Chapter 2 can be used to produce summary statistics for individual cells of variables other than those used for cross-tabulation. The subcommands for this purpose are MEANS(mean), MEDIANS(median), SUMS(sum), MINIMUM, MAXIMUM, STDEV(standard deviation), and STATS(count, mean, and standard deviation). Such statistical subcommands require an argument, specifying the column or variable of data on which the computation is to be made.

1. In the following example, the CODE commands are used to collapse 'Age' and 'Work_hrs' into two categories first. Then the TABLE command with the COUNTS and MEANS subcommands tabulates the number of observations and the average GPA for students who do and do not work more than 15 hours, and for those who are and are not older than 22. These are the marginal values. In addition, Counts and GPA averages are given per cell for all four cells. Thus there are 4 students who are under 22 and who work 15 hours or less. Their average GPA is 2.8625. Then there are another 4 students of the same age but who work 16 hours or more. Their average GPA is higher: 3.3450. In fact the same results hold for those over 22 years old.

```
MTB > CODE (18:22)1 (23:50)2 'Age' C20
MTB > CODE (0:15)1 (16:40)2 'Work_hrs' C21
MTB > TABLE C20 by C21;
SUBC> COUNTS;
SUBC> MEANS 'GPA'.

ROWS: C20     COLUMNS: C21

              1         2        ALL

    1         4         4          8
          2.8625    3.3450     3.1037

    2         1         3          4
          2.5400    3.0000     2.8850

  ALL         5         7         12
          2.7980    3.1971     3.0308

    CELL CONTENTS --
                    COUNT
                GPA:MEAN
```

3.4 COMPREHENSIVE SUMMARY OF DATA

3.4.1 Descriptive Summay of Data: The DESCRIBE Command

Thus far we have dealt with <u>one</u> statistic at a time such as the mean or the standard deviation. Frequently we wish to obtain various statistics for variables all at once instead of entering the commands one by one. The DESCRIBE command provides such comprehensive statistics for specified columns.

Syntax:
```
DESCRIBE the data in C...C;
     BY   C
```

This command provides the following summary statistics for each column specified:

N	Number of nonmissing observations
N*	Number of missing observations
MEAN	Mean
MEDIAN	Median
TRMEAN	A 5% trimmed mean, after removing the smallest 5% and the largest 5% of the values.
STDEV	Standard deviation
SEMEAN	Standard error of the mean
MIN	Minimum
MAX	Maximum
Q1	First quartile (25th percentile)
Q3	Third quartile (75th percentile)

The subcommand, BY, groups the data according to the categories of the variable listed after BY, and computes summary statistics for each category.

1. Let's obtain descriptive statistics for final exam scores.

```
MTB > DESCRIBE 'Final'

                N      MEAN    MEDIAN   TRMEAN    STDEV   SEMEAN
Final          12     74.75     75.00    74.70    10.00     2.89

               MIN       MAX        Q1        Q3
Final        60.00     90.00     65.25     84.25
```

2. We will now compare the final exam scores of males and females. Since 'Sex' was entered as alpha data, it must be converted into numeric data before DESCRIBE with BY can be performed.

CONVERT converts 'Sex' with alpha data into numerical codes, and stores the result in C13. A conversion table is generated first, assigning a value of 1 to males and 2 to females. Refer to Section 3.5.3 which explains the CONVERT command in greater detail. DESCRIBE displays descriptive statistics for males and females separately.

```
MTB > READ C11-C12;
SUBC> FORMAT(A1,1X,F1.0).
DATA> M 1
DATA> F 2
DATA> END
MTB > CONVERT C11 C12 'Sex' C13
MTB > NAME C13 'N_Sex'
MTB > DESCRIBE 'Final';
SUBC> BY 'N_Sex'.
```

	N_Sex	N	MEAN	MEDIAN	TRMEAN	STDEV	SEMEAN
Final	1	5	77.00	75.00	77.00	10.37	4.64
	2	7	73.14	75.00	73.14	10.22	3.86

	N_SEX	MIN	MAX	Q1	Q3
Final	1	65.00	90.00	67.50	87.50
	2	60.00	89.00	65.00	82.00

The output shows that in this sample the mean and the median for men are higher than those for women. Dispersion of scores, measured by the standard deviation, the range (MAX-MIN), and the interquartile (Q3-Q1) is smaller for men than for women.

3.5 COLUMN AND ROW MANIPULATION

While describing statistically the characteristics of our data, we can benefit from some commands which manipulate column and row entries. This section presents those which create new columns from the existing columns by copying, stacking, or unstacking. Also, commands dealing with alpha data are discussed.

3.5.1 Copying Columns: The COPY Command

Syntax:
```
COPY from C ... C into C ... C;
    USE    rows K ... K
    USE    rows where C = K ... K
    OMIT   rows K ... K
    OMIT   rows where C = K ... K

COPY from constants K ... K into K ... K
COPY from C into constants K ... K
COPY from constants K ... K into C
```

The COPY command copies data from columns to columns, from constants to constants, from columns to constants, or from constants to columns.

The subcommand USE or OMIT selectively copies a portion of the data.

1. The following command copies stored constants in K1, K2, and K3 into Column 5(C5).

```
MTB > RESTART
MTB > LET K1=1
MTB > LET K2=2
MTB > LET K3=3
MTB > COPY K1-K3 C5
MTB > PRINT K1-K3 C5
K1      1.00000
K2      2.00000
K3      3.00000

C5
   1    2    3
```

2. The next example copies rows 5 through 10 of C1 to C10.

```
MTB > SET C1
DATA> 1:10
DATA> END
MTB>  COPY C1 C10;
SUBC> OMIT 1:4.
MTB > PRINT C10
C10
   5    6    7    8    9   10
```

3. Let's retrieve the class data, and create two new columns of final exam scores according to the reasons for taking this course. These columns should differentiate those who are taking the course because of necessity (Reason=1,4) and those who are interested in the subject (Reason=3). Afterwards, compute the means for these subgroups.

```
MTB > RETRIEVE 'A:CLASS'
MTB > COPY 'Final' C11;
SUBC> USE 'Reason'=1 4.
MTB > NAME C11 'Forced'
MTB > COPY 'Final' C13;
SUBC> USE 'Reason'=3.
MTB > NAME C13 'Interest'
MTB > PRINT C1 'Final' 'Reason' 'Forced' 'Interest'
```

ROW	Name	Final	Reason	Forced	Interest
1	Abbot,John	90	3	82	90
2	Baker,Kathy	89	3	60	89
3	Cook,Dianne	82	1	65	85
4	Hayes,Fred	60	1	70	
5	Lind,Mary	75	2	75	
6	Moore,Tom	65	1	66	
7	O'Connor,Steve	70	1	65	
8	Ray,Alex	85	3	75	
9	Smith,Dick	75	4		
10	Conrad,Jane	66	4		
11	Hardy,Jill	65	4		
12	Freund,May	75	1		

```
MTB > MEAN of 'Forced'
    MEAN    =      69.750
MTB > MEAN OF 'Interest'
    MEAN    =      88.000
```

The final exam scores of those who are taking the course because of their interest in the subject are 90, 89, and 85 with the mean score of 88.0. In contrast there are 8 students who are enrolled out of necessity, whose mean score is 69.75.

3.5.2 Stacking Columns: The STACK Command

While COPY duplicates a portion of or the whole column, the STACK command stacks a block (a number of columns) of numeric or alpha data and/or constants on top of another block or other blocks. This command is useful in creating a composite column from data contained in different columns.

Syntax:

```
STACK  (E ... E)... on top of (E...E) put in (C...)
    SUBSCRIPTS   into C
```

Under the STACK command the first block is placed on top of the second block. Each block is enclosed in parentheses. Thus the first column of the first block is stacked on top of the first column of the second block. Therefore each block must contain the same number of columns. Columns within each block must be of the same length, i.e., with the same number of rows. When the blocks consist of one column only, the parentheses may be omitted.

To keep track of the original location, the SUBSCRIPTS subcommand creates a new column, which contains values indicating which block each row came from. The value 1 in the subscript column refers to the rows of the destination column which is derived from the first block, while a value of 2 is from block 2, and so on.

STACK can be used for both numeric and alpha data, although SUBSCRIPTS must be numeric.

1. In the following example C1 is stacked on top of C3, then on top of C3 again. The result is stored in C6. C7 contains C5 topped by C4, then topped by C2.

```
MTB > READ C1-C5
DATA> 1 2 3 4 5
DATA> 1 2 3 4 5
DATA> 1 2 3 4 5
DATA> END
MTB > STACK (C1 C2) (C3 C4) (C3 C5) (C6 C7);
SUBC> SUBSCRIPTS C8.
```

Let's see how these numbers are stacked up by printing all eight columns. Examine the subscript values in C8 carefully.

```
MTB > PRINT C1-C8
  ROW   C1   C2   C3   C4   C5   C6   C7   C8
```

ROW	C1	C2	C3	C4	C5	C6	C7	C8
1	1	2	3	4	5	1	2	1
2	1	2	3	4	5	1	2	1
3	1	2	3	4	5	1	2	1
4						3	4	2
5						3	4	2
6						3	4	2
7						3	5	3
8						3	5	3
9						3	5	3

2. A simpler example is illustrated by stacking scores of 'Exam1' on top of the scores of 'Exam2' in the class data, then stored in C15. Note that the parentheses are omitted around 'Exam1' and 'Exam2' since there is only one column in each block. Since the previous example has replaced 'Exam1' by C8 in the worksheet, you need to retrieve the data again.

```
MTB > RETRIEVE 'A:CLASS'
MTB > STACK 'Exam1' on 'Exam2' store in C15
MTB > NAME C15 'All'
MTB > PRINT 'Exam1' 'Exam2' 'All'
```

53

ROW	Exam1	Exam2	All
1	90	95	90
2	100	92	100
3	80	77	80
4	65	60	65
5	76	82	76
6	62	70	62
7	80	*	80
8	90	88	90
9	70	65	70
10	88	75	88
11	67	73	67
12	65	77	65
13			95
14			92
15			77
16			60
17			82
18			70
19			*
20			88
21			65
22			75
23			73
24			77

3.5.3 Unstacking Columns: <u>The UNSTACK Command</u>

The reverse of STACK is the UNSTACK command, which breaks down one block of columns into several smaller blocks of columns or constants.

Syntax:
```
UNSTACK  (C...C) into (E...E)...(E...)
   SUBSCRIPTS  are in C
```

The SUBSCRIPTS subcommand is necessary most of the time because it designates the column which contains values to separate the original blocks of columns into their component blocks. Rows with the smallest subscript value are placed in the first block, rows with the second smallest value go into the second block, and so on. The subscripts must be integers from -1000 to +10000 or missing values (*).

All blocks must be enclosed in parentheses unless there is only one column involved.

1.   Let's group the final exam scores into separate columns according to the reasons why students are taking this course. The subscript values therefore are provided by 'Reason'. This task can be performed by the COPY command used earlier, but

54

UNSTACK is more efficient if the intent is to break down columns according to the subscript values.

```
MTB > UNSTACK ('Final') into (C21) (C22) (C23) (C24);
SUBC> SUBSCRIPT in 'Reason'.
MTB > NAME C21 '1' C22 '2' C23 '3' C24 '4'
MTB > PRINT 'Final' 'Reason' C21-C24
```

ROW	Final	Reason	1	2	3	4
1	90	3	82	75	90	75
2	89	3	60		89	66
3	82	1	65		85	65
4	60	1	70			
5	75	2	75			
6	65	1				
7	70	1				
8	85	3				
9	75	4				
10	66	4				
11	65	4				
12	75	1				

3.5.4 Converting Data Between Alpha and Numeric Formats: The CONVERT Command

For a categorical variable such as 'Reason' numerical codes (1,2,3..) have been assigned. Sometimes we wish to display more descriptive labels such as 'Forced', 'Interest', and the like. Conversely, there are occasions where we need to transform alpha data into numerical codes so that statistical analysis can be performed. The CONVERT command accomplishes such conversion of alpha data to numeric data or vice versa.

Syntax:
```
CONVERT using conversion table in C,C, convert C to C
```

Before using CONVERT, a conversion table must be furnished to Minitab, assigning a numeric value to each alpha variable.

1. Let's convert the numeric variable in C7 'Reason' into alpha data. First, create a conversion table using C11 and C12. A format subcommand is needed to read into the conversion table because one column contains alpha data. C11 with a (F1.0) format holds the numerical codes, while C12 with an (A8) format contains the alpha labels. Using this conversion table in C11-C12, the data in the 'Reason' column are altered and stored in C13.

```
MTB > READ C11-C12;
SUBC> FORMAT(F1.0,1X,A8).
DATA> 1 FORCED
DATA> 2 TEACHER
DATA> 3 INTEREST
DATA> 4 TIME
DATA> 5 OTHER
DATA> END
MTB > CONVERT using conversion table in C11 C12, &
CONT>            convert 'Reason' to C13
MTB > PRINT 'Name' 'Reason' C13
```

ROW	Name	Reason	C13
1	Abbot,John	3	INTEREST
2	Baker,Kathy	3	INTEREST
3	Cook,Dianne	1	FORCED
4	Hayes,Fred	1	FORCED
5	Lind,Mary	2	TEACHER
6	Moore,Tom	1	FORCED
7	O'Connor,Steve	1	FORCED
8	Ray,Alex	3	INTEREST
9	Smith,Dick	4	TIME
10	Conrad,Jane	4	TIME
11	Hardy,Jill	4	TIME
12	Freund,May	1	FORCED

3.5.5 Concatenating Alpha Data: <u>The CONCATENATE Command</u>

In displaying alpha data there is another command which is useful. The CONCATENATE command links together alpha columns to generate a single column.

Syntax:

```
CONCATENATE  C,...,C put into C
```

After the command keyword CONCATENATE, a list of columns to be pasted together follows. However, unlike most Minitab commands such as COPY, the storage column (the last column on the command) under CONCATENATE must be different from the input columns specified.

1. Suppose the last name is entered in C1, and the first name initial is in C2. Let's change the format so that the first name initial is followed by a period and then the last name.

    RESTART is executed to wipe out the worksheet labeled, CLASS.MTW. The last name and the first name initial of 4 observations are read into C1 and C2 as specified by the format.　Four dots are created by SET, and are kept in C4. CONCATENATE the first name initial (C2), the period (C4), and the last name (C1) in that order, and store the result in C5.

```
MTB > RESTART
MTB > READ C1-C2;
SUBC> FORMAT(A10,A1).
DATA> Abbot     J
DATA> Baker     K
DATA> Cook      D
DATA> Hayes     F
DATA> END
MTB > SET C4;
SUBC> FORMAT(A1).
DATA> .
DATA> .
DATA> .
DATA> .
DATA> END
MTB > CONCATENATE C2 C4 C1 in C5
MTB > PRINT C5

C5
 J.Abbot        K.Baker        D.Cook         F.Hayes
```

```
NEW MINITAB COMMANDS

   Arithmetic functions, Column-wise & Row-wise statistics
   LET
   DESCRIBE
   COPY
   STACK, UNSTACK
   CONVERT, CONCATENATE
```

EXERCISES

1. Classify students into three groups ('Cl_load') by the number of units carried.　Read only necessary columns from the data file.

2. Compute the mean and standard deviation of GPA for the three groups of 'Cl_load', using various Minitab commands such as:

   a. TABLE; BY.
   b. COPY;USE or STACK, and column MEAN.

CHAPTER FOUR
GRAPHING DATA

```
In This Chapter You Will Learn To:

     1.   Save Minitab sessions and commands
     2.   Use DOS system commands within Minitab
     3.   Execute batch mode Minitab
     4.   Graph one to K-Sample data
     5.   Plot two to K-variable relations
     6.   Read from an external data file
```

So far you have learned to tabulate frequency distributions and compute statistics to describe the characteristics of sample data. This chapter presents another method of summarizing the data: the use of graphs. Graphs create a striking visual impact upon the reader, allowing an immediate grasp of the shape of the distribution.

Before discussing graphic presentation, however, this chapter first deals with the relationship between a Minitab session and the DOS environment. Such knowledge is vital for efficient execution of Minitab commands.

4.1 THE MINITAB AND DOS OPERATING SYSTEM

4.1.1 Saving the Minitab Session: The OUTFILE and PAPER Commands

You have learned how to save data in a worksheet format (SAVE) or in an ASCII format (WRITE). Now you will learn how to save the Minitab session itself for viewing. The OUTFILE command stores all of the commands as well as the Minitab responses in a file with the extension LIS. This file is in ASCII format so that you can read and edit it with a word processor to produce a report. You can also extract Minitab commands from this file, and execute it in a batch mode. (See Section 4.1.4)

Syntax:
```
OUTFILE   'filename'

     OW = K
     OH = K
     NOTERM
```

The OW subcommand sets the output width of the printer or file, while the OH subcommand sets the page size for the output.

The NOTERM subcommand stops sending the output to the terminal, directing it to the printer or an output file.

To terminate recording Minitab session and close this LIS file, type the NOOUTFILE command:

1. Let's start copying the current Minitab session onto a file, called TASK1.LIS, in drive A. Specify the file name without the extension and drive in parentheses after OUTFILE.

    ```
    MTB > OUTFILE 'A:TASK1'
    ```

2. Everything will be stored in this file, TASK1.LIS, until you execute the following command:

    ```
    MTB > NOOUTFILE
    ```

The PAPER Command

The PAPER command allows you to send all or a portion of the Minitab session to a printer.

Syntax:
```
PAPER

    OW = K
    OH = K
    NOTERM
```

The subcommands are used in the same way as those for OUTFILE.

1. To start printing your session, type:

    ```
    MTB> PAPER
    ```

    To stop the process, issue the command:

    ```
    MTB> NOPAPER
    ```

2. You cannot use the OUTFILE command and the PAPER command simultaneously. Issuing the PAPER command while OUTFILE is in effect will deactivate OUTFILE. The reverse is also true.

4.1.2 Saving Commands: The JOURNAL Command

While OUTFILE stores both input (commands) and output (results) lines, the JOURNAL command copies only command lines to a journal file with the extension MTJ. Unlike PAPER, this command can be used with OUTFILE at the same time. This file can be executed in a batch mode, which will be discussed in Section 4.1.4.

```
Syntax:   JOURNAL  ['filename']
          NOJOURNAL
```

NOJOURNAL stops copying commands onto a journal file.

1. Let's save the commands from this point on. We will retrieve the data in a worksheet, CLASS.MTW, and run TABLE and DESCRIBE commands on the final examination scores.

```
MTB > JOURNAL 'A:PROG1'
MTB > RETRIEVE 'A:CLASS'
MTB > TABLE 'Final'

 ROWS: Final

    COUNT

60       1
65       2
66       1
70       1
75       3
82       1
85       1
89       1
90       1
ALL     12

MTB > DESCRIBE 'FINAL'
```

	N	MEAN	MEDIAN	TRMEAN	STDEV	SEMEAN
Final	12	74.75	75.00	74.70	10.00	2.89

	MIN	MAX	Q1	Q3
Final	60.00	90.00	65.25	84.25

4.1.3 DOS System Commands

Now that a Minitab session and commands are stored in files in ASCII format, it will be convenient if we could invoke DOS commands from within the Minitab session. There are certain DOS commands which can be executed without leaving Minitab. They are CD, DIR, and TYPE.

1.   Let us find out what files are stored in drive A by DIR.

```
MTB > DIR A:
A:\

CLASS.DAT    CLASS.MTW    PROG1.MTJ    TASK1.LIS
```

We see the names of the files which we have created. The file CLASS.DAT was created by the WRITE command, CLASS.MTW was created by SAVE, PROG1.MTJ by JOURNAL, and TASK1.LIS by OUTFILE.

2.   Since CLASS.DAT is written in ASCII, we can display its contents by TYPE as follows:

```
MTB > TYPE A:CLASS.DAT

    Abbot,John      M   19   1   3.50   20   3    90   95   90
    Baker,Kathy     F   20   2   3.68   25   3   100   92   89
    Cook,Dianne     F   21   3   2.90   18   1    80   77   82
    . . .
    Freund,May      F   20   2   2.95    2   1    65   77   75
```

The SYSTEM Command

In addition, there is the SYSTEM command which allows you temporarily to move out to the DOS environment without stopping a Minitab session. Just type SYSTEM at the MTB> prompt, then the Minitab session is suspended and you are in the DOS system. You can execute DOS commands at the DOS prompt. When you are done, type EXIT to return to Minitab.

1.   Let's go into DOS by typing SYSTEM.

```
MTB > SYSTEM
```

2.   Now let's print the data file. The DOS command, PRINT, works in the same way as in a regular DOS session. It outputs the file on paper through the printer.

```
C:\MINITAB > PRINT A:CLASS.DAT
Name of list device [PRN]: LPT1
```

3.   Now let's type the journal file.

```
C:\MINITAB > TYPE A:PROG1.MTJ
```

```
RETRIEVE 'A:CLASS'
TABLE 'FINAL'
DESCRIBE 'FINAL'
DIR A:
TYPE A:\CLASS.DAT
(Then follows the output of CLASS.DAT since it is printed in
DOS, not MINITAB)
SYSTEM
```

Using the Editor with which you are familiar, edit PROG1.MTJ to read as follows:

```
RETRIEVE 'A:CLASS'
TALLY 'FINAL';
ALL.
DESCRIBE 'FINAL';
BY 'MAJOR'.
```

Save the above commands in a file called PROG1.MTB. Note that it is the extension MTB which makes the file executable in a batch mode. You will learn in the next section how to execute this file, containing Minitab commands.

4.   Type EXIT to return to the previous Minitab session.

4.1.4 The Batch Mode Operation: The EXECUTE Command

An interactive mode is useful because you receive an instantaneous response from the computer for each line you type in. A disadvantage is, however, that you cannot go back to the previous commands to make changes. In a batch mode operation you can write, edit, and store a series of commands in a file, and execute these commands altogether.

The EXECUTE command executes commands stored in a command file. A command file can be created by (1) JOURNAL as described in the previous section, (2) editing a file created by OUTFILE, (3) the macro command STORE which will be discussed in Chapter 5, or (4) a text editor such as EDLIN.

Syntax: | EXECUTE   commands [in 'filename'] [K times] |

EXECUTE is followed by the name, the path and drive, if necessary, of a command file in single quotation marks. Make sure that the command file has the extension of MTB. If no filename is specified, Minitab will execute a file, MINITAB.MTB, which is created by a macro facility (See Chapter 5).

You may execute the same batch of commands more than once by specifying the desired number of repetitions.

1. Let us execute PROG1.MTB created and edited in the previous section (4.1.3). Note that the file name extension is not entered inside the single quotation marks.

```
MTB > EXECUTE 'A:PROG1'
```

2. When PROG1.MTB is executed, the Minitab commands, messages, and output will be displayed on the screen as follows:

```
MTB > RETRIEVE 'A:CLASS'
  WORKSHEET SAVED  8/28/1990

Worksheet retrieved from file: A:CLASS.MTW
MTB > TALLY 'Final';
SUBC> ALL.

   Final  COUNT CUMCNT PERCENT  CUMPCT
      60      1      1    8.33    8.33
       .  .  .
```

3. You may want to enter the NOJOURNAL and NOOUTFILE commands now to stop the recording of your output and commands in TASK1.LIS and PROG1.MTJ.

4.1.5 A Summary of File Types

We have learned to create various kinds of files. Let us summarize these files by their filename extensions, the tasks they perform, and the commands that create them.

Command	File extension	Description
OUTFILE NOOUTFILE	LIS	Minitab session (commands and outputs) stored in an ASCII format.
SAVE RETRIEVE	MTW	Worksheet saved as a binary file, which can be input by RETRIEVE.
SAVE;PORTABLE RETRIEVE;PORTABLE	MTP	Portable worksheet saved, which can be input with RETRIEVE with PORTABLE.
WRITE READ,INSERT,SET	DAT	ASCII data file from the worksheet, which can be input with READ, SET, and INSERT
STORE EXECUTE	MTB	ASCII file containing macro commands, which is executed by EXECUTE.
JOURNAL NOJOURNAL	MTJ	ASCII file of commands saved from a Minitab session.

4.2 UNIVARIATE PLOTTING: ONE OR K SAMPLE DATA

From tabulating and running descriptive statistics, we know that the students in our class data are young. However, the question of "how young?" is probably answered more intuitively by looking at a graphic representation of the age distribution. Minitab provides various plotting devices to portray a distribution. In this section, you will learn to use these devices to graph age.

4.2.1 Histogram: The HISTOGRAM Command

The HISTOGRAM command draws a histogram of data in columns together with the midpoint and count of values.

Syntax:
```
HISTOGRAM  of the data in C...C
   INCREMENT = K
   START      at K [end at K]
   BY         C
   SAME       scales for all columns
```

The optional subcommand INCREMENT specifies the width, K, of each interval, while the subcommand START indicates the midpoint for the first (and the last) interval.

The subcommand BY breaks down the data into groups according to the values in column C, and draws a histogram for each group.

With the subcommand, SAME, Minitab uses a common scale for all columns listed on the HISTOGRAM command.

1. Let us draw histograms to examine the age distribution by sex, using HISTOGRAM. Since we are going to look at only two variables, we will read two columns from the data file instead of retrieving the entire file. We have created a data file named CLASS.DAT in Chapter 2 with the WRITE command and have not accessed it up to this point, the reason being that it is much easier to retrieve a worksheet stored through the SAVE command as long as we are working with MINITAB. However, WRITE does have the advantage of creating a file in ASCII which is accessible and readable through other software programs.

    The READ command with a file specification reads data from an external data file in an ASCII format with the DAT extension. In other words, it is the appropriate command used for retrieving a file created by the WRITE command.

Since 'Sex' was stored as alpha data, READ must be accompanied by a format subcommand. As you may recall, in the data file 'Sex' is stored in C2, and 'Age' is in C3. If CLASS.DAT was created by WRITE without a FORMAT subcommand, you should type the data file out in an editor program to confirm the locations in which these variables are found. Because of one space for the carriage control before C1, 14 spaces used for the longest name, and one space between C1 and C2, there are 16 spaces before 'Sex' is identified. 'Sex' takes the format (A1) because it is represented by one character. Following 'Sex', 2 spaces are skipped before 'Age' is found. If a FORMAT subcommand is used with WRITE, then just READ the file using the same format.

The data file, CLASS.DAT, does not store the column names. Therefore you must insert names again after retrieving the data into the current worksheet. The following example illustrates the retrieval of the data file which has been written without a format subcommand. Additionally, it creates a histogram of 'Age' regardless of 'Sex'.

```
MTB > READ 'A:CLASS' C2-C3;
SUBC> FORMAT(16X,A1,2X,F2.0).
MTB > NAME C2 'Sex' C3 'Age'
MTB > HISTOGRAM of 'Age'

 Histogram of Age    N = 12

 Midpoint    Count
       20        5    *****
       22        3    ***
       24        0
       26        2    **
       28        0
       30        0
       32        1    *
       34        0
       36        0
       38        0
       40        0
       42        1    *
```

2. In the above example, MINITAB chooses an interval of 2 for the histogram. In this example, let us use a 5-year interval in drawing the same histogram. We could also specify the midpoint of the first interval as 20 with the INCREMENT and START subcommands. As shown below, students are heavily concentrated in the age bracket 18-22 years.

```
MTB > HISTOGRAM OF 'Age';
SUBC> INCREMENT=5;
SUBC> START AT 20.
```

```
Histogram of Age    N = 12

Midpoint    Count
   20.00        8    ********
   25.00        2    **
   30.00        1    *
   35.00        0
   40.00        1    *
```

3. Next, let's compare the age distributions of males and females in the class. The subcommand BY of HISTOGRAM takes only a numeric variable or column. Therefore we must convert 'Sex' into numeric codes called 'N-sex'. You have learned how to use the CONVERT command to accomplish this in the last chapter. Recall that a conversion table is created first by reading alpha and numeric values into C11 and C12. Using this table, histograms are drawn for males (N_sex=1) and females (N_sex=2) as specified.

```
MTB > READ C11-C12;
SUBC> FORMAT(A1,1X,F1.0).
DATA> M 1
DATA> F 2
DATA> END
MTB > CONVERT using C11 C12, convert 'Sex' into C13
MTB > NAME C13 'N_sex'
MTB > HISTOGRAM OF 'Age';
SUBC> INCREMENT=5;
SUBC> START AT 20;
SUBC> BY 'N_sex'.
```

```
Histogram of Age    N_sex = 1    N = 7

Midpoint    Count
   20.00        4    ****
   25.00        1    *
   30.00        0
   35.00        0
   40.00        0


Histogram of Age    N_sex = 2    N = 5

Midpoint    Count
   20.00        4    ****
   25.00        1    *
   30.00        1    *
   35.00        0
   40.00        1    *
```

4.2.2 Horizontal Histogram: The DOTPLOT Command

Minitab provides another type of histogram, which plots horizontally with the DOTPLOT command. DOTPLOT divides the X-axis into more divisions, resulting in a more detailed picture of the distribution. It is useful for small data sets such as the one we have on hand. For a larger data set, HISTOGRAM shows the general shape better.

Syntax:

```
DOTPLOT  of the data in C...C
    INCREMENT    = K
    START        at K [end at K]
    BY           C
    SAME         scale for all columns
```

The subcommands are used in the same manner as those of HISTOGRAM.

1. Let's draw a horizontal histogram of 'Age'.

```
MTB > DOTPLOT OF 'Age'


                  .
         : : : .          :                   .                      .
  -------+---------+---------+---------+---------+---------+---------Age

      20.0      25.0      30.0      35.0      40.0      45.0
```

2. DOTPLOT is better than HISTOGRAM when comparing two or more groups because the former displays the plots for various groups, stacked vertically for easier visual examination. Note that the variable 'N_sex' was created in the last illustration of HISTOGRAM.

```
MTB > DOTPLOT OF 'Age';
SUBC> BY 'N_sex'.
```

```
N_sex
1
             :     .   .          .
  -------+---------+---------+---------+---------+---------+---------age

N_sex         .
2          : .           .               .                   .
  -------+---------+---------+---------+---------+---------+---------age

      20.0      25.0      30.0      35.0      40.0      45.0
```

4.2.3 Stem and Leaf Plot: The STEM-AND-LEAF Command

The STEM-AND-LEAF command also constructs a histogram, but it uses digits representing actual values rather than symbols.

Syntax:

```
STEM-AND-LEAF   of the data in C...C
    TRIM        outliers
    INCREMENT   = K
    BY          C
```

This command produces three columns. The leftmost column presents a cumulative count of values from the top down to the middle value (median), and from the bottom up to the middle value, but not including the count of the median interval. The median is the value in the middle. The frequency or count of the interval containing the median is enclosed in parentheses. Parentheses are omitted if the median falls between two rows in the display.

The second column represents a stem digit with its leaf digits shown in the third column. The leaf unit indicates the location of the decimal point in each number. Thus a stem of 24 and a leaf of 6 with a leaf unit of 1.0 would represent the number 246. If the leaf unit is 0.10, the number is 24.6. Sometimes the leaves for one stem are stretched over more than one line of the display.

The TRIM subcommand excludes outliers from the stem-and-leaf plot, and shows them on lines labeled LO and HI. The subcommands TRIM and BY cannot be used simultaneously.

The INCREMENT subcommand specifies an incremental value in vertical scaling. The difference between the smallest values in adjacent rows is indicated by K.

The BY subcommand generates a stem-and-leaf plot for each subgroup identified by the values in C.

1.   Let's draw a stem-and-leaf plot of 'Age'.

```
MTB > STEM-AND-LEAF of 'Age'

 Stem-and-leaf of Age      N  = 12
 Leaf Unit = 1.0

     2     1 99
    (6)    2 000112
     4     2 55
     2     3 2
     1     3
     1     4 1
```

According to the output there are 2 cases in the teens: two 19 year old students. Between 20 and 24 there are six students aged 20, 20, 20, 21, 21, and 22. Since the median is found within this interval the cumulation of cases is done up to but not including this interval. Note also the bracketing of the frequencies or counts at this interval to indicate the location of the median. From the other end of the scale, there is one student in the 40's, namely, age 41. No case falls in the range 35-39, and one student, age 32, falls in the range 30-34. The first column shows the count of 2 in the third row from the bottom because of the two students aged 41 and 32. Then in the next interval, there are 2 more students, aged 25, thus providing a cumulative count of 4. The next category of 20-24 is where the median is located.

2. Male and female comparison is made by using the subcommand BY.

```
MTB > STEM-AND-LEAF of 'Age';
SUBC> by 'N_sex'.

  Stem-and-leaf of Age        N_sex = 1        N  = 7
  Leaf Unit = 1.0

      2      1 99
     (2)     2 12
      1      2 5


  Stem-and-leaf of Age        N_sex = 2        N  = 5
  Leaf Unit = 1.0

     (4)     2 0001
      3      2 5
      2      3 2
      1      3
      1      4 1
```

4.2.4 Box-and-Whisker Plot: <u>The BOXPLOT Command</u>

Instead of plotting the actual values, a box plot displays summary statistics for the distribution. It plots the median, the 25th and the 75th percentiles, and extreme cases which are called outliers.

Syntax:

```
BOXPLOT    for data in C
    START      at K [end at K]
    INCREMENT = K
    NOTCH      [K% confidence] sign confidence interval
    BY         C
    LEVELS     K...K [for C]
    LINES      =K
    FILE       'filename'
```

The BOXPLOT command produces a box and whisker display. The graph consists of a rectangular box with two dashed lines (whiskers), extending in opposite direction from two sides of the box called hinges because of the way in which they are drawn. The box represents the middle half of the data, which covers the first to the third quartile of the data. The median is marked by a + which appears within the box. The first and third quartiles are demarcated by the hinges ("I") as described previously. From this graph, a number of statistics can be defined as follows:

(H-spread) = (upper hinge - lower hinge)
inner fences = (lower hinge) - 1.5*(H-spread) and (upper hinge) + 1.5*(H-spread)
outer fences = (lower hinge) - 3*(H-spread) and (upper hinge) + 3*(H-spread)

Whiskers extend from the hinges of the box to the corresponding adjacent values. Outliers are the values between the inner and outer fences, which are represented by asterisks. Values beyond the outer fences are called extreme outliers, and are plotted with an O.

The START and INCREMENT subcommands control the scaling of the horizontal axis. START specifies the first (and the last) tick mark (+ symbol) on the axis, while INCREMENT determines the interval between the tick marks.

The NOTCH subcommand produces notches (parentheses) in a boxplot, representing the confidence interval for the population median. Unless K% is specified, the default is a 90% confidence interval.

The BY subcommand produces separate boxplots for subgroups classified by the values in C. When you wish to produce boxplots for selected subgroups or rearrange the order of the subgroup displays, use the LEVELS subcommand. If the BY variable has 5 categories (e.g., 1: White 2: Black 3: Hispanics 4: Asian 5: Other), and you wish to produce boxplots for Asians and blacks only, the LEVEL subcommand is necessary as follows:

```
MTB > BOXPLOTS FOR C1;
SUBC> BY 'RACE';
SUBC> LEVELS 4 2.
```

Boxplots are normally displayed in three lines. The option LINES = 1 condenses the plot horizontally from three lines to one.

The FILE subcommand stores the graphics output in a specified file.

1. Let's draw a box plot for 'Age'. The plot shows that 50% of the students fall in the age bracket 20-25, and that the median is between 21 and 22. The whiskers stretch long into the early 30s but short into the teens. There is an extreme outlier in the 40s.

70

```
MTB > BOXPLOT 'Age'
```

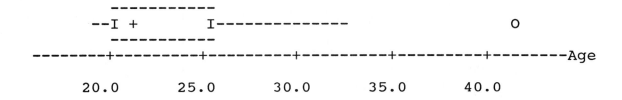

2. In addition, we will compare the box plots of males and females, and also indicate 90% confidence intervals by NOTCH. Since we want a comparison of males and females, the BY 'N_sex' subcommand is necessary. To make the visual comparison easier, let's present two plots close to each other by setting LINES=1.

```
MTB > BOXPLOT 'Age';
SUBC> NOTCH;
SUBC> BY 'N_Sex';
SUBC> LINES=1.

N_Sex

 1            (   + I---)--
 2          ( +              I-------)----------------
      --------+---------+---------+---------+---------+--------Age
           20.0      25.0      30.0      35.0      40.0
```

4.3 BIVARIATE PLOTTING

Thus far we have plotted the distribution of only one variable. Our next step is to display the relationship between two variables graphically. For example, you would like to know if GPA is related to the number of hours students work per week. The PLOT command produces a scatterplot of two variables.

4.3.1 Plotting a Two Variable Relationship: <u>The PLOT Command</u>

The PLOT command plots the data in the first column (vertical axis) against the data in the second column (horizontal axis). Each point is represented by an *, unless the SYMBOL subcommand changes it to some other symbol. If two or more points overlap, a count is printed instead of an *. When more than 9 points fall on the same location, the + symbol is displayed.

```
Syntax:    PLOT        C versus C
              YINCREMENT   = K
              YSTART       at K [end at K]
              XINCREMENT   = K
              XSTART       at K [end at K]
              SYMBOL       'symbol'
              TITLE    'text'
              FOOTNOTE 'text'
              YLABEL   'text'
              XLABEL   'text'
```

The size of the PLOT can be controlled by the HEIGHT and WEIGHT commands, which will be discussed in Section 4.3.2.

The YINCREMENT subcommand specifies the interval on the vertical axis with K indicating the distance between the tick marks (+). The YSTART subcommand controls the first (and the last) value(s) to be displayed.

The XINCREMENT and XSTART subcommands act in a similar manner for the horizontal axis.

Four labeling subcommands make the plot more easily readable. It is possible to specify up to three TITLE subcommands with one plot. TITLES are centered above the plot. You may specify up to two FOOTNOTES with one plot, which are left-justified below the plot. One YLABEL and one XLABEL may be assigned with each plot. The y-axis label, printed vertically, is centered along the y-axis. The x-axis label is centered under the x-axis.

1.   Let us plot 'GPA' on the y-axis against 'Work_hrs' on the x-axis. Since we may be using other variables, we retrieve the entire class worksheet file.

```
MTB > RETRIEVE 'A:CLASS'
MTB > PLOT 'GPA' vs 'Work_hrs';
SUBC> TITLE 'GPA versus Weekly Hours of Work';
SUBC> YLABEL 'GPA';
SUBC> XLABEL 'Working hours'.
```

Not surprisingly as we have learned from the last chapter, the plot shows that students who spend more hours at work maintain higher GPA than those who work less. However, the relationship is not linear.

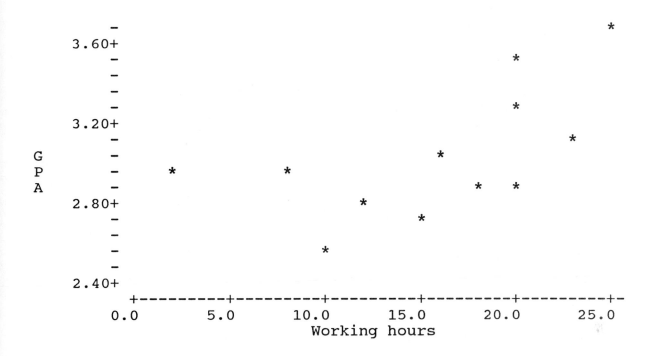

2. PLOT can be used to draw an ogive (cumulative frequency distribution) of a variable such as the final exam score. First sort 'Final' in ascending order, and store it as 'Final-S' in C12. Count the number of final exam scores, and store it in K1. Generate integers from 1 to the total number of cases (12) and store them in C13. The relative cumulative frequency is obtained by dividing each one of these integers by the total number of cases and storing them in C14. Plot the relative cumulative frequency against the final exam scores which are sorted in ascending order.

```
MTB > SORT 'Final' C12
MTB > NAME C12 'Final-S'
MTB > COUNT 'Final' K1

      COUNT    =       12

MTB > SET C13
DATA> 1:K1
DATA> END
MTB > DIVIDE C13 by K1 put in C14
MTB > NAME C14 'RLFRQ'
MTB > PLOT 'RLFRQ' by 'Final-S';
SUBC> YLABEL 'RLFRQ';
SUBC> XLABEL 'Final-S'.
```

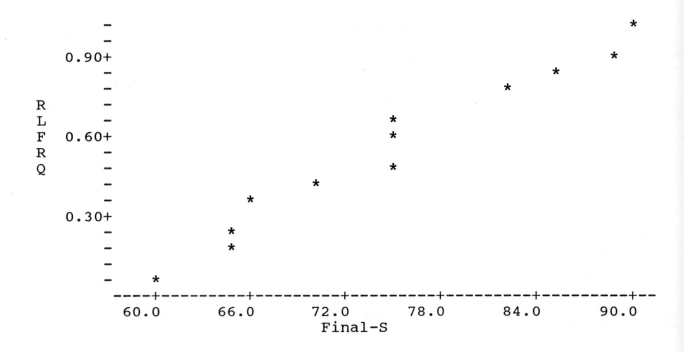

4.3.2 Control of Plot Size: The HEIGHT and WIDTH Commands

There are Minitab commands which control the size of a plot. The HEIGHT command controls the height of PLOT, MPLOT, LPLOT, TPLOT, and TSPLOT. It has no effect on the height of HISTOGRAM, DOTPLOT, STEM-AND-LEAF, or BOXPLOT.

Syntax: | HEIGHT   of all plots that follow is K lines |

The range of K values is from 5 to 400 lines, although the use of an odd number tends to produce better scaling. The default height is 17 lines.

The WIDTH command controls the width of BOXPLOT, DOTPLOT, LPLOT, MPLOT, PLOT, and TPLOT. WIDTH ranges from 10 to 150 spaces. The default value is 57 spaces.

Syntax: | WIDTH   of all plots that follow is K spaces |

4.4 MULTIVARIATE PLOTTING

4.4.1 Overlaying Scatterplots: The MPLOT Command

Relationships among three variables can be examined by overlaying two-dimensional scatterplots on top of one another. For example, you may wonder if the reasons for taking the course affect the first exam and the final exam scores differently. One way to answer this question is by superimposing a plot of 'Final' against 'Reason' upon another plot of 'Exam1' against 'Reason'. This is accomplished by the MPLOT command.

MPLOT overlays multiple pairs of X-Y coordinates, representing two columns each on the same set of axes. Instead of *'s, the first pair of columns is identified by the symbol A, the second by B, and so on. If more than one point fall on the same spot, a count is printed. For an explanation of the subcommands, see the section on the PLOT command.

Syntax:

```
MPLOT           C vs C and C vs C and ... C vs C
    YINCREMENT     = K
    YSTART         at K [end at K]
    XINCREMENT     = K
    XSTART         at K [end at K]
    TITLE     'text'
    FOOTNOTE  'text'
    YLABEL    'text'
    XLABEL    'text'
```

1. Let us overlay two scatter plots to examine differential effects of 'Reason' upon 'Final' and 'Exam1'. The x-axis is set to start at 0 with an interval of 1 so that the values of 'Reason' will be displayed at 1, 2, 3, and 4.

For reason 2 (the teacher is appealing), there is no visible difference between the first and the final exam scores if we assume the 2 represents an A and a B. Among students with reasons 3 (interest in the subject) and 4 (time convenience), the first exam scores are higher than the final exam scores. That is, they did better in the beginning of the semester than later. It is difficult to tell, because of the overlay, what the distribution of B is like for reason 1. For example, the count of '2' could have been derived from an A and a B, or 2 A's, or 2 B's. However, in general, those who took the course because of their interest in the subject (Reason=3) received higher scores both on 'Exam1' and 'Final'.

```
MTB > MPLOT 'Final' vs 'Reason' and 'Exam1' vs 'Reason';
SUBC> YINCREMENT=20;
SUBC> YSTART AT 40;
SUBC> XINCREMENT=1;
SUBC> XSTART AT 0;
SUBC> TITLE 'Final & Exam 1 against Reason';
SUBC> YLABEL 'Exams';
SUBC> XLABEL 'Reason'.
```

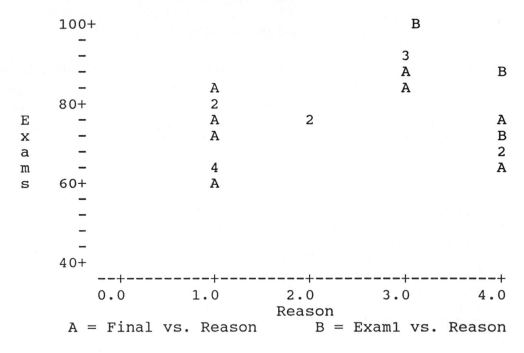

4.4.2 Subgrouping Scatterplots: <u>The LPLOT Command</u>

Another way of viewing the relationship among the three variables is to use the LPLOT command. It produces scatterplots for different groups, and overlays them on one another using the same axes. Thus the relationship between 'Exam1' and 'Final' can be scatterplotted for four subgroups based on 'Reason'.

Syntax:
```
LPLOT      C vs C, groups in C
   YINCREMENT     = K
   YSTART         at K [end at K]
   XINCREMENT     = K
   XSTART         at K [end at K]
   TITLE      'text'
   FOOTNOTE   'text'
   YLABEL     'text'
   XLABEL     'text'
```

According to this command, the first column (Y-axis) is plotted against the second column (X-axis), denoting points with letters (A,B,C,...) corresponding to the subgroups in the third column. An (X,Y) pair for group 1 is plotted with an "A", the second group with a "B", and so on.

For an explanation of the subcommands, see the PLOT command section above.

1. In the example below 'Final' is plotted against 'Exam1' for each subgroup created by 'Reason'. Scatter points representing the first group (Reason=1) are marked by "A"s, the points for the second group (Reason=2) are marked by "B"s, the third group by "C"s, and the fourth group by "D"s.

```
MTB > LPLOT 'Final' vs 'Exam1' in 'Reason';
SUBC> YLABEL 'Final';
SUBC> XLABEL 'Exam 1'.
```

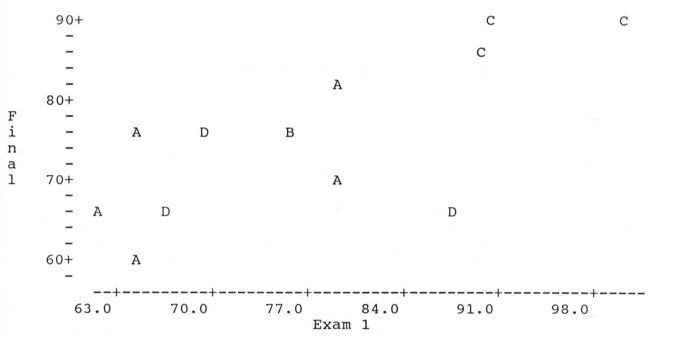

The overall relationship between 'Final' and 'Exam1' is far from linear, but the tendency is for students who have high scores on one test to have high scores also on the other test. Those who are taking the course because of their interest in the subject (Reason=3, marked by a "C") are found at the upper right corner, i.e., high scores on both 'Final' and 'Exam1'. The other three groups ("A", "B", "D") are scattered in the lower score region, but no special subgroup pattern is discernible.

4.4.3 Three Dimensional Plot: The TPLOT Command

Finally we will examine the three dimensional plot for displaying the relationship among three variables. The TPLOT command produces a pseudo three dimensional plot of X-Y-Z. It is "pseudo" because the third dimension is indicated by symbols representing the values of Z.

```
Syntax:    TPLOT   y in C vs x in C vs z in C

               YINCREMENT = K
               YSTART at K [go to K]
               XINCREMENT = K
               XSTART at K [go to K]
               TITLE      'text'
               FOOTNOTE 'text'
               YLABEL     'text'
               XLABEL     'text'
```

The following symbols are used, where zbar is the mean of the numbers in the Z column, and s is its standard deviation.

"0" represents:	z is 1 standard deviation below the mean, z < zbar-s
"." represents:	z is between 1 standard deviation below the mean and the mean itself, zbar-s < z < zbar
"/" represents:	z is between 1 standard deviation above the mean and the mean itself, zbar < z < zbar+s
"X" represents:	z is 1 standard deviation above the mean, z > zbar+s

If several points fall on the same spot, a count is given as in PLOT.

The subcommands, YINCREMENT, YSTART, XINCREMENT, XSTART, work in the same way as in the PLOT command.

To control the size of TPLOT, use the WIDTH and HEIGHT commands.

1.   Let us examine the relationship among the final exam score, reasons for taking the course, and the weekly number of hours at work. Use 'Final' as the Y-axis, 'Reason' as the X-axis, and 'Work_hrs' as the pseudo Z-axis.

The mean of 'Work_hrs' is 15.75 and the standard deviation is 6.70. Among those who are taking the course because of requirement (Reason=1), those who work between 9.05 and 15.75 hours a week (as represented by ".") have the lowest final exam scores (below 70). Those who work less than 9.05 hours (as represented by "0"), and others who work between 15.75 and 22.45 hours (as represented by "/") have higher scores (in the 70's and in the 70's and 80's).

78

Those who work more than 22.45 hours ("X") are taking the course because of their liking of the teacher, and their final exam scores are in the 70s.

We will leave the rest of the interpretation to you. Since it is a pseudo three dimensional plot, the visual effect is not very straightforward. However, it does offer detailed information.

```
MTB > TPLOT 'Final' vs 'Reason' vs 'Work_hrs';
SUBC> XINCREMENT=1;
SUBC> XSTART 0.
```

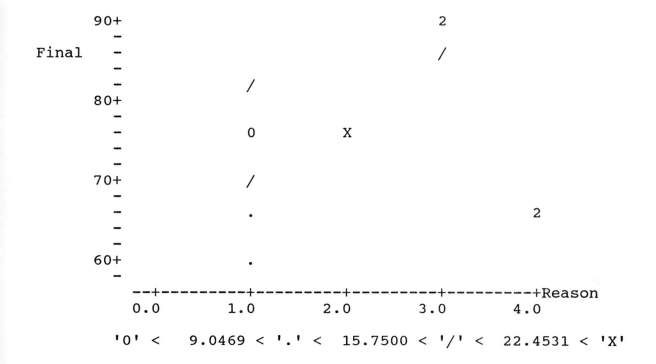

There are many other Minitab commands for graphic representation, including CONTOUR, time series plotting, and high resolution graphics. Due to space limitation, however, they will not be discussed in this book. If you are interested, use the Help facility to explore the use of these commands.

```
NEW MINITAB COMMANDS

    OUTFILE, NOOUTFILE, PAPER, NOPAPER
    JOURNAL, NOJOURNAL
    DOS commands: CD, DIR, TYPE
    EXECUTE
    HISTOGRAM, DOTPLOT, STEM-AND-LEAF, BOXPLOT
    HEIGHT, WIDTH
    PLOT, MPLOT, LPLOT, TPLOT
    COUNT
```

EXERCISES

1. Draw histograms and boxplots for the number of working hours by the different class levels.

2. Draw a scatterplot for the number of units against the number of working hours per week.

3. Study the relationships among GPA, the level of satisfaction with college life, and the marital status by using MPLOT, LPLOT, and TPLOT. Before plotting, it is necessary to generate integer values for each variable by rounding or coding the original values.

CHAPTER FIVE
PROBABILITY AND SAMPLING DISTRIBUTION

```
In This Chapter You Will Learn To:

    1.   Use probability functions such as
           PDF: Probability density function
           CDF: Cumulative distribution function
           INVCD:Critical values and percentiles of
           distribution
    2.   Generate sampling distributions
    3.   Simulate the Central Limit Theorem
    4.   Use the Macro facility
```

Up to this point our attention has been focused on descriptive statistics. Having described and summarized the characteristics of samples, the next step is to infer properties of populations on the basis of sample results. Such inferential statistics are based on probability theory. This chapter discusses many useful Minitab commands which can be used to enhance the understanding of the concepts of probability and sampling distributions. These concepts provide a foundation for hypothesis testing and estimation, which are covered in the remaining chapters of this book.

5.1 THE CONCEPT OF PROBABILITY

5.1.1 Probability as Chance

"Should I buy stock in the XYZ corporation today or wait for a few more days?" Pete wonders. As a result of the Persian Gulf crisis, the price may go down further, or it may come back. The chances seem to him to be 50-50, so he tosses a coin to decide. Minitab can simulate this situation in case no coin is available.

```
MTB > NOTE    Buy: 1    Don't buy: 0
MTB > RANDOM 1 toss, put into C1;
SUBC> BERNOULLI trials with p=.5.
MTB > PRINT C1
  C1
      1
```

In the above example, he decides to buy the stock if a head appears, and not to buy if a tail appears. The command RANDOM with the subcommand BERNOULLI generates a random value (1 or 0) for Bernoulli trials. As you know, a Bernoulli trial consists of two possible outcomes.

The result is then stored in C1. When he prints C1, he knows what he should do. According to this trial, he will buy the stock because C1 prints "1". For further explanation of the RANDOM command, see Section 5.6.

Jim is trying to decide whether he should sell his house now or next year. Based on the real estate situation in his community, he thinks that there is one chance in 6 that the house will appreciate significantly during the coming year. So he lets the cast of a die decide for him. If "1" turns up, he will hold on to the house for another year. Again he uses Minitab to do the job because he does not have a die at hand.

This time he must use the subcommand INTEGER with the command RANDOM. INTEGER generates random observations from a discrete uniform distribution on specified ranges such as 1-6. In other words, the INTEGER subcommand is used when there are more than 2 outcomes, with the same probability (uniform distribution) attached to each outcome.

The result he obtained is 2, so he will sell his house now.

```
MTB > RANDOM 1 roll, put into C1;
SUBC> INTEGER from 1 to 6.
MTB > PRINT C1
C1
      2
```

If Jim rolls the die six times, would he get each value once? Let's try and see. Specify the desired number of observations (6) after the command keyword, RANDOM. As you can see, although the probability of each value is the same, he obtains "1" three times and no "4" or "6".

```
MTB > RANDOM 6 rolls, put into C2;
SUBC> INTEGER 1 6.
MTB > PRINT C2
C2
      5   1   2   3   1   1
```

5.1.2 Probabilities as Long-Term Relative Frequencies:
(The Law of Large Numbers)

As you see in the previous section, you normally do not get the values 1 to 6 once each when you roll a die six times. The probability of an event happening, i.e., a specific number that comes up on a die, must be viewed as the relative frequency of its occurrence in the long run as opposed to a single or a few rolls. Suppose you toss a coin 100 times and get 48 heads. Then the relative frequency is 48/100=0.48. The relative frequency of heads is likely to approach 0.5 if you toss a fair coin a very large number of times. Probability is therefore viewed as the long-term relative frequencies of repeated experiments which stabilize after a certain number of repeats.

Minitab is able to simulate an experiment based on a large number of repetitions, producing random outcomes rapidly and accurately. Let us simulate the tossing of a coin 50 times. After each toss, calculate the ratio of the number of heads thus far obtained against the number of tosses made so far. Then plot the ratio (relative frequency) against the number of trials.

```
MTB > NOTE Law of Large Numbers
MTB > RANDOM 50 obs in C1;
SUBC> BERNOULLI p=0.5.
MTB > SET C3
MTB > 1:50
MTB > END
MTB > PARSUMS C1, put in C2
MTB > DIVIDE C2 by C3, put in C4        #Successive ratios
MTB > NAME C3 'Obs'  C4 'Ratio'
MTB > PLOT ratios in C4 vs No of tosses in C3

Ratio
      -
      -                              *********
 0.60+        *                    ****            **
      -          *              *  **                 ***
      -    * * * * * * ** *                          ***              *
      -                    *                            *********
      -              * *
 0.40+
      -
      -      *
      -
      -
 0.20+
      -
      -
      -
      -
-0.00+  *
        +---------+---------+---------+---------+---------+-Obs
        0        10        20        30        40        50

MTB > PRINT C1-C4
   ROW    C1    C2    Obs       Ratio

     1     0     0     1    0.000000
     2     1     1     2    0.500000
     3     0     1     3    0.333333
     4     1     2     4    0.500000
     5     1     3     5    0.600000
     6     0     3     6    0.500000
     7     1     4     7    0.571429
     8     0     4     8    0.500000
     . . .
```

83

The PARSUMS command computes the partial sums of elements in C1, and puts the results into C2. In the above example for the first row, C1 = 0, therefore C2=0. In the second row C1=1. Add the values of C1 in row 1 and 2 to obtain PARSUMS; therefore C2 = 0 + 1 = 1. In the third row C2 = 0 + 1 + 0 = 1.

The number of trials for each row is entered by the SET command. To produce integers from 1 to 50, type 1:50 at the DATA prompt.

The ratios of success(C4) are computed by the LET command where the partial sum of successes(C2) is divided by the number of trials(C3) up to that point.

The relative frequency of success(C4) is plotted against the number of trials(C3). The plot shows the relative stability of the ratio converging to 0.5, as the number of trials increases.

5.2 PROBABILITY OF EVENTS

An event E is a subset of the outcome space S. As such, the probability of an event P(E) is the sum of the probabilities for all those outcomes in E.

The subcommand DISCRETE under the RANDOM command produces random observations from a finite sample space S = $\{e_1, e_2, ... e_k\}$ where the probabilities $P(e_i)$, i=1,2,...k are assumed to be known. The DISCRETE subcommand, unlike INTEGER, allows different probabilities to be associated with various outcomes.

Imagine a letter sent from San Francisco to Tokyo. Let the outcome space S={3,4,5,6,7} be the number of days it takes for the letter to be delivered. Assume P(3)=0.1 P(4)=0.3 P(5)=0.3 P(6)=0.2 P(7)=0.1. Simulate the number of days required to deliver 1000 letters, and compute the relative frequencies with which a letter arrives in 3, 4, 5, 6, or 7 days.

The DISCRETE subcommand under the RANDOM command generates random values based on a discrete distribution with specified values and their associated probabilities.

In the following example the number of days(C1) and their probabilities(C2) are entered by SET commands. 1000 random values are generated into C3 from the discrete distribution with C1 values and their C2 probabilities.

```
MTB > SET C1                             # Outcome space
DATA> 3 4 5 6 7
MTB > END
MTB > SET C2                             # Probability associated
DATA> 0.1   0.3   0.3   0.2   0.1
MTB > END
```

```
MTB > RANDOM 1000 C3;
SUBC> DISCRETE C1 C2.
MTB > HISTOGRAM C3

 Histogram of C3    N = 1000
 Each * represents 10 obs.

 Midpoint     Count
        3        86   *********
        4       284   ****************************
        5       302   ******************************
        6       211   *********************
        7       117   ************
```

The INDICATOR Command

The INDICATOR command can be used to create dummy variables (indicators), representing each category of the column(C3), that is, 3-day, 4-day,... delivery time. A dummy variable has values of 1 or 0, where 1 symbolizes the occurrence and 0, the non-occurrence of the category.

Syntax:

```
INDICATOR variables for levels of C, put into C,...,C
```

Consider the following example:

```
MTB > INDICATOR for C3, put in C4-C8
MTB > NAME C4 '3-DAY' C5 '4-DAY' C6 '5-DAY' C7 '6-DAY'
MTB > NAME C8 '7-DAY'
MTB > PRINT C3-C8
  ROW    C3   3-DAY 4-DAY 5-DAY 6-DAY 7-DAY

    1     4      0     1     0     0     0
    2     7      0     0     0     0     1
    3     5      0     0     1     0     0
    4     4      0     1     0     0     0
    . . .

MTB > MEAN C4
    MEAN    =      0.086000
MTB > MEAN C5
    MEAN    =       0.28400
MTB > MEAN C6
    MEAN    =       0.30200
MTB > MEAN C7
    MEAN    =       0.21100
MTB > MEAN C8
    MEAN    =       0.11700
```

If there are K categories/levels for a variable, then K storage columns must be created. The storage columns will contain 1 in the specific column in which the occurrence of the category/level is noted, and a 0 in all other columns for that row.

In the above example, the first and the fourth rows of the levels column (C3) show a 4, which is the second category/level (where K=3,4,5,6,7). Thus the second storage column (C5) contains 1 in the first and fourth rows. Other columns (C4, C6-C8) of the first and the fourth rows are 0s.

The means of these indicator variables show the relative frequencies of 3- 4- 5- 6- and 7-day delivery time in the 1000 simulated trials to be 0.086, 0.284, 0.302, 0.2111, and 0.117.

5.3 PROBABILITY DISTRIBUTIONS

Minitab can replace the statistics tables at the end of your textbook by displaying probabilities and cumulative probabilities for a given value in a given distribution. The PDF command gives the probability of X successes, and the CDF command gives the probability of X successes or fewer in n trials in discrete distributions. For continuous distribution (chisquare, F, normal, t, and uniform), PDF calculates the continuous probability density function, and gives the height of the curve for each value specified.

The INVCDF command generates an inverse cumulative distribution function, which can be used to obtain critical values and percentiles of a given distributions. PDF, CDF, and INVCDF have the same syntax as shown below:

```
Syntax:     PDF     for values in E [store results in E]
            CDF     for values in E [store results in E]
            INVCDF for values in E [store results in E]

            Subcommands:
              BETA            EXPONENTIAL   LOGISTIC      UNIFORM
              BINOMIAL        F             LOGNORMAL     WEIBULL
              CAUCHY          GAMMA         NORMAL
              CHISQUARE       INTEGERS      POISSON
              DISCRETE        LAPLACE       T
```

After the command keyword, specify the value(s) for which you wish to have PDF, CDF or INVCDF calculated. Also indicate the location for output.

For selected subcommands, arguments are shown below:

BINOMIAL	n = K p = K
POISSON	mean = K
INTEGER	discrete uniform on K to K
DISCRETE	values in C, probabilities in C
NORMAL	[mu = K, [sigma = K]]
UNIFORM	[continuous on K to K]
T	degrees of freedom = K
F	numerator df = K denominator df = K
CHISQUARE	df = K

5.3.1 Binomial Distribution

If you wish to know the probability of getting 2 heads in tossing a coin 4 times, the following Minitab session will give you the answer.

```
MTB > NOTE Probability of 2 heads in 4 tosses of a coin
MTB > PDF of the value 2 , store in K1;
SUBC> BINOMIAL n=4 p=0.5.
MTB > PRINT K1
 K1          0.375000
```

In order to obtain the entire set of probabilities for all possible values in n trials, do not specify a value after the PDF command. The following Minitab command creates a probability table for binomial distribution where n=20 and p=0.2.

```
MTB > PDF;
SUBC> BINOMIAL n=20 p=0.2.

        BINOMIAL WITH n =   20  p = 0.200000
            K            P( X = K)
            0              0.0115
            1              0.0576
            2              0.1369
            3              0.2054
            4              0.2182
            5              0.1746
            6              0.1091
            7              0.0545
            8              0.0222
            9              0.0074
           10              0.0020
           11              0.0005
           12              0.0001
           13              0.0000
```

In order to output the probabilities for further processing, two specifications are necessary after the PDF command. The first refers to the values, while the second the output probabilities. In the following example, the values needed (0-20) are created by the SET command into C1 and their corresponding probabilities are stored into C2 by PDF. The probabilities (C2) are then plotted against these values (C1).

```
MTB > SET C1
DATA> 0:20
MTB > END
MTB > PDF of the values in C1 store in C2;
SUBC> BINOMIAL n=20 p=.2.
MTB > PLOT C2 VS C1
```

```
C2      -
        -
        -
        -                    *
0.210+              *
        -
        -
        -           *
        -
0.140+        *
        -
        -              *
        -
        -
0.070+
        -    *          *
        -
        -         *
        - *              *
-0.000+              *   *  *   * *   * *   * *  * *
        +---------+---------+---------+---------+---------+-C1
       -0.0      4.0       8.0      12.0      16.0      20.0
```

The cumulative density function is obtained by the CDF command as follows:

```
MTB > CDF;
SUBC> BINOMIAL n=20 p=0.2.

BINOMIAL WITH N =   20   P = 0.200000
   K   P( X LESS OR = K)
   0          0.0115
   1          0.0692
   2          0.2061
   3          0.4114

   . . .
  10          0.9994
  11          0.9999
  12          1.0000
```

The INVCDF command generates the inverse cumulative distribution function. The following example shows the inverse relationship between CDF and INVCDF. According to this illustration, the INVCDF returns a value X, i.e., 1.96, such that an observed value less than or equal to X has the user-specified probability of E, i.e., E=0.975, under the distribution defined in the subcommand, which is a normal distribution with a mean of 0 and a standard deviation of 1. The CDF command on the other hand takes a user-specified X value, 1.96 in this case, and returns a cumulative probability, .975, for X equal to or less than 1.96.

```
MTB > CDF 1.96;
SUBC> NORMAL 0 1.
        1.96      0.9750
MTB > INVCDF 0.975;
SUBC> NORMAL 0 1.
        0.975      1.9600
```

In the case of the binomial distribution you may not get an exact INVCDF value. With the specification of 0.3 for INVCDF below, the two closest K values are selected for output although neither gives the exact probability of 0.3.

```
MTB > INVCDF 0.3;
SUBC> BINOMIAL 100 0.2.
       K   P(X LESS OR = K)          K   P(X LESS OR = K)
       17             0.2712         18             0.3621

MTB > CDF 18;
SUBC> BINOMIAL 100 0.2.
       K   P( X LESS OR = K)
       18             0.3621
```

5.3.2 Poisson Distribution

Suppose the average number of telephone calls arriving in a 5-minute period is 5. Generate the probability density function for the Poisson distribution.

```
MTB > PDF;
SUBC> POISSON mean=5.

POISSON WITH MEAN =    5.000
    K                P( X = K)
    0                  0.0067
    1                  0.0337
    2                  0.0842
    3                  0.1404
    . . .
    14                 0.0005
    15                 0.0002
    16                 0.0000
```

To plot the above probabilities, integers from 0 through 16 are created by the SET command, and are stored in C1. The PDF of the Poisson distribution for the values in C1 is stored in C2. PLOT plots the probability density function for X=0 to 16.

```
MTB > SET C1
DATA> 0:16
MTB > END
MTB > PDF C1 C2;
SUBC> POISSON mean=5.
MTB > PLOT C2 vs C1
```

```
C2      -                    *    *
        -
        -
0.150+                          *
        -           *
        -
        -
        -
0.100+                              *
        -
        -      *
        -                            *
        -
0.050+
        -                              *
        -    *
        -                                *
        -  *                          *
0.000+                          *  *   *  *  *
        +---------+---------+---------+---------+---------+--C1
      -0.0       3.0       6.0       9.0      12.0      15.0
```

The cumulative density function for the Poisson distribution with mean=5 is plotted in the following example. The CDF for values in C1 (from the previous example) is generated and stored in C3 first. Then the cumulative probabilities in C3 are plotted against the values 1 through 16 in C1.

```
MTB > CDF C1 C3;
SUBC> POISSON mean=5.
MTB > PLOT C3 vs C1
```

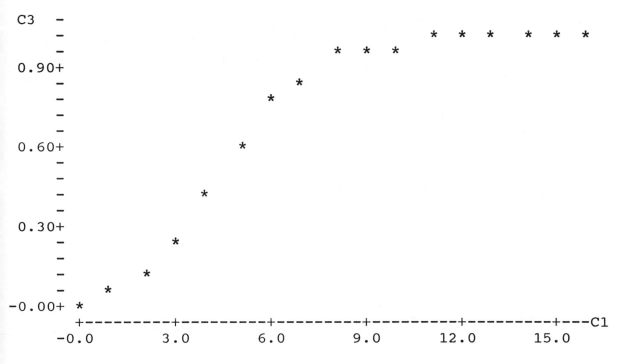

5.3.3 Normal Distribution

If we wish to know the ordinate (height) of a value of a random variable with a normal distribution and a mean of 0 and standard deviation of 1, we would proceed as follows. At x=1:

```
MTB > PDF at 1.0;
SUBC> NORMAL mu=0 sigma=1.
      1.00     0.2420
```

What is the cumulative probability at Z=1.96?

```
MTB > CDF at 1.96;
SUBC> NORMAL mu=0 sigma=1.
      1.96     0.9750
```

Let us plot the normal distribution curve with mean=0 and standard deviation=1. The data entry in the SET command below instructs Minitab to produce numbers ranging from -4 to +4 with an interval of 0.2.

```
MTB > SET C1
DATA> -4:4/.2
MTB > END
MTB > PDF at C1, put in C2;
SUBC> NORMAL mu=0 sigma=1.
MTB > PLOT C2 vs C1
```

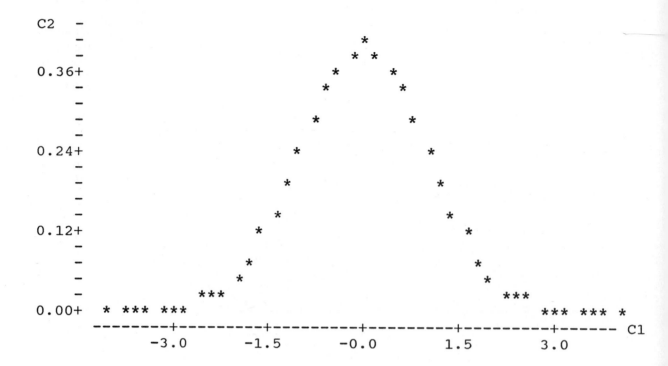

5.4 SAMPLING DISTRIBUTIONS

There are two Minitab commands which are used to generate samples: RANDOM and SAMPLE. The RANDOM command generates a random sample from a specified distribution with replacement. The SAMPLE command, on the other hand, selects rows of data at random from a worksheet. SAMPLE with the subcommand REPLACE draws a sample with replacement. Without the subcommand, SAMPLE produces a sample without replacement. For further elaboration, see Section 5.6.

5.4.1 Sampling from a Known Probability Distribution

To generate a random sample with two values, use the BERNOULLI subcommand with p=0.5:

```
MTB > RANDOM 10 observations, put in C1;
SUBC> BERNOULLI p=0.5.
MTB > PRINT C1
 C1
     1    1    1    1    1    0    1    1    1    1
```

To generate a random sample with more than two values, use the INTEGER subcommand. As you recall, INTEGER generates a random sample from a discrete uniform distribution on a specified interval (1-6).

```
MTB > RANDOM 10 observations, put in C1;
SUBC> INTEGER 1 6.
MTB > PRINT C1
 C1
      6    5    3    6    4    1    4    6    1    5
```

To generate a sample of values with unequal probabilities, use the DISCRETE subcommand. In the following example, $P(1)=0.4$, $P(2)=0.5$, $P(3)=0.1$.

```
MTB > READ values in C1 probabilities in C2
DATA> 1   0.4
DATA> 2   0.5
DATA> 3   0.1
MTB > END
MTB > RANDOM 10 observation with values 1,2,or 3, put in C3;
SUBC> DISCRETE C1 C2.
MTB > PRINT C3
 C3
      2    1    2    2    3    2    2    2    1    1
```

5.4.2 Sampling from a Finite Population

To draw a random sample from a finite population stored in a worksheet, you must use the SAMPLE command. SAMPLE randomly selects a specified number of rows from a worksheet. Without the subcommand, REPLACE, sampling is done without replacement. With the subcommand, REPLACE, a sample is drawn with replacement. In sampling without replacement, an observation/value, once drawn for a sample, is not available for selection again and the population is reduced by 1. On the other hand, an observation can be selected again and again in sampling with replacement, since the observation is returned to the population after it is sampled.

```
MTB > NOTE Sampling without replacement
MTB > SET C1          #Create a population with values
DATA> 1:10            # 1,2,3,4,5,6,7,8,9,10
DATA> END
MTB > SAMPLE 5 obs from C1, put in C6   # Without replacement
MTB > SAMPLE 5 obs in C7;               # With replacement
SUBC> REPLACE.
MTB > NAME C6 'Without' C7 'With'
MTB > PRINT C6 C7
  ROW  Without   With

   1        8       2
   2        5       8
   3       10       9
   4        3      10
   5        2      10
```

93

In the above example, a sample of size 5 is drawn first from a population consisting of integers 1 through 10 <u>without</u> replacement. The second sample is drawn <u>with</u> replacement using the REPLACE subcommand. As the printout shows, the first sample contains 5 different values, while the second sample has the same number (10) appearing twice.

In an actual survey it is necessary to sample from several columns at once. In the following example the data set contains student's ID, GPA, and the number of study hours. A sample of three students with their ID, GPA, and study hours out of the 6 students is drawn without replacement, and the results are put in new columns.

```
MTB > READ C1-C4
DATA> 1   21   3.55   20
DATA> 2   18   2.58   10
DATA> 3   20   3.05   15
DATA> 4   29   3.59   28
DATA> 5   20   3.34   19
DATA> 6   19   2.84   17
DATA> END
MTB > SAMPLE 3 from C1-C4, put into C11-C14
MTB > PRINT C11-C14
  ROW    C11    C12    C13    C14

    1      6     19   2.84     17
    2      3     20   3.05     15
    3      2     18   2.58     10
```

5.4.3 Stratified Sampling

Let us consider stratified sampling. After the worksheet is stratified by 'Sex', a proportionate sample is drawn from each gender. The purpose is to compare the income of males and females.

```
MTB > NOTE Stratify proportionately by Sex
MTB > NOTE Sample size = 100
MTB > READ 'A:SURVEY' C1 C2          # Enter population
MTB > NAME C1 'Sex' C2 'Income'      # Two variables entered
MTB > UNSTACK C2 into C3-C4;         # Break down population
MTB > SUBSCRIPT C1.                  #   into males & females
MTB > NAME C3 'Males' C4 'Females'
MTB > LET K1 = COUNT(C3)/COUNT(C1)   # Proportion of males
MTB > LET K2 = COUNT(C4)/COUNT(C1)   # Proportion of females
MTB > LET K3 = ROUND(100*K1)         # Male sample size
MTB > LET K4 = ROUND(100*K2)         # Female sample size
MTB > SAMPLE K3 from C3, put into C5 # Male sample
MTB > SAMPLE K4 from C4, put into C6 # Female sample
```

First, read 'Sex' and 'Income' from a study population in the data file, SURVEY.DAT. Using UNSTACK, divide the population into males (C3) and females (C4). To find the proportion of males in the population, divide the number of males, COUNT(C3) by the size of the population, COUNT(C1).

Since the desired sample is 100, multiply 100 by the proportion of males to obtain the required number of males in the sample (K3). Since the K3 and K4 values should be integers in the following SAMPLE commands, to be on the safe side their values are rounded off to the closest integer with the ROUND function. SAMPLE the required number of males (K3) from the column (C3) containing the incomes of males. Do the same for females.

5.5 THE CENTRAL LIMIT THEOREM

A remarkable feature of Minitab is the ease and speed with which it can simulate repeated sampling. Let us simulate repeated sampling to illustrate the Central Limit Theorem. The Central Limit Theorem states: When random samples of size n are taken from any population with a finite mean μ and a finite standard deviation σ, the resulting sampling distribution of sample means approximates a normal distribution with a finite mean μ and a standard deviation of $\sigma/\sqrt{n}$ if n is large. An important point is that the population from which samples are taken need not be normally distributed, but can assume almost any form.

5.5.1 Population Distribution

We will use a non-normal population from which samples are to be taken. The population has a discrete uniform distribution with values ranging from 1 to 6. The mean is 3.5 and the standard deviation is 1.8708. The flat shape of the distribution is seen by plotting the probability density function (PDF) against the values 1 through 6.

```
MTB > SET C1
DATA> 1:6
MTB > END
MTB > PDF C1 C10;
SUBC> INTEGER 1 6.
MTB > MEAN C1
    MEAN      =       3.5000
MTB > STDEV C1
    ST.DEV.  =       1.8708
MTB > PLOT C10 vs C1;
SUBC> YINCREMENT 0.1;
SUBC> YSTART 0 END 0.3.
```

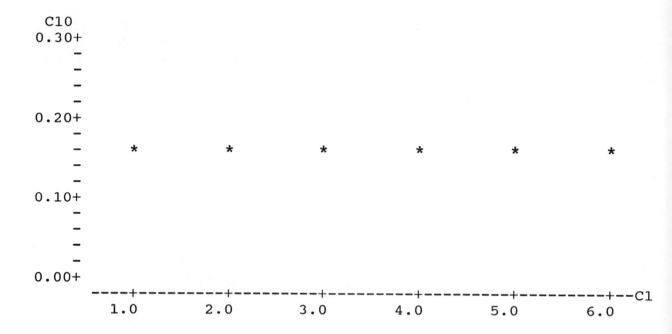

5.5.2 Simulation 1: Repeated Sampling

To simulate the Central Limit Theorem we can take one sample at a time to compute its mean. After repeating this process many times, we can obtain a sampling distribution of sample means.

Let us begin with a sample of 10 rolls of a die. What is the mean score? The following Minitab commands simulate this experiment, producing the sample mean of 3.382.

```
MTB > NOTE Method 1 -- Repetitious simulation
MTB > RANDOM 10 obs in C1;
SUBC> INTEGER 1 6.
MTB > MEAN of C1
   MEAN = 3.382
```

Repeating the above process, 40 times for example, becomes extremely tedious. Furthermore, it is necessary to store a sample mean each time so that the mean of these sample means can be computed later.

5.5.3 Simulation 2: Row-Wise Technique

To avoid this repetition, the data can be arranged differently by entering the values of each sample row-wise instead of column-wise.

In the following example the first 40 random numbers, representing the outcomes from the rolls of a die, are stored in C1. The next 40 outcomes will be put in C2, and so on. After drawing 10 samples, we have a worksheet containing 40 rows and 10 columns of random values with the range of 1 through 6. Since they are all randomly generated, we can interpret the first row as consisting of a sample of size 10. The second row is the second sample, again of size 10. Thus we have 40 samples of size 10. Computation of the sample mean is done by first adding the columns, C1-C10 in a row, then dividing the sums in C11 by 10. The mean and the standard deviation of the 40 sample means(C12) are computed by MEAN and STDEV.

```
MTB > RANDOM 40 obs in C1-C10;
SUBC> INTEGER 1 6.
MTB > PRINT C1-C10
  ROW    C1    C2    C3    C4    C5    C6    C7    C8    C9   C10

    1     6     2     4     6     3     6     3     6     2     3
    2     5     6     4     6     5     3     6     5     1     5
    3     2     6     1     4     6     2     2     6     1     3
    4     2     6     1     6     2     3     1     2     1     3
    5     3     5     1     2     2     6     2     2     1     4

   . . .

MTB > ADD C1-C10 put in C11          # Sample size: 10
MTB > DIVIDE C11 BY 10 put in C12    # C12 : Sample means
MTB > MEAN C12                       # Mean,40 sample means
     MEAN      =       3.5925
MTB > STDEV C12
     ST.DEV. =       0.45482
MTB > HISTOGRAM C12

  Histogram of C12    N = 40

Midpoint       Count
     2.8           1   *
     3.0           4   ****
     3.2           2   **
     3.4          10   **********
     3.6           8   ********
     3.8           5   *****
     4.0           3   ***
     4.2           3   ***
     4.4           3   ***
     4.6           0
     4.8           1   *
```

5.5.4 Simulation 3: The Macro Technique

The above technique generates samples very fast but cannot handle a large sample size and many repetitions because of the limitations of the worksheet. As the third method, we therefore use a <u>Macro</u> to facilitate simulation.

A Macro is a sequence of commands stored in a macro file. Once a macro is created, you can issue these commands all at once by simply executing one command, EXECUTE 'filename'. When you need to execute a block of commands repeatedly, specify the number of times the execution is to take place after the filename. Thus, the macro utility makes the repeated processing of a group of commands easy and fast, without user entry of these commands every time.

A macro begins with the STORE command, and ends with the END command. All the commands between STORE and END will be stored in the file specified after STORE. They are entered at the STORE> prompt. The NOECHO command suppresses the display of the commands in a macro when the macro is executed. Without it, the commands will be shown on the screen. The ECHO command resumes the display of commands.

In the example below STORE opens a file, MACRO3.MTB, in drive B to store commands. The file extension MTB is automatically attached to MACRO3 by Minitab. At the STOR> prompt, commands to be stored are entered. There are two arguments K1 and K2 in the file, whose values will be assigned at the time of execution. K1 stands for the number of observations to be generated by RANDOM. K2 represents the row number for the column C3. The mean of values in C2 is computed and assigned to the K2'th position of C3. After the K2'th position is filled, K2 is incremented by 1 to point to the next location of C3.

After defining the macro, execute it with properly assigned values. Since the sample size is 10, K1 is 10. K2 should be initialized to be 1 to point to the first row of C3.

A macro can be executed a specified number of times, which in this case is 40. After repeating this macro 40 times, C3 will contain 40 sample means from samples of size 10.

The BASE command initializes the seed value to 1000 for generating random numbers. Without this command, RANDOM produces a different set of numbers each time it is run.

```
MTB > STORE 'B:MACRO3'
STOR> NOECHO
STOR> RANDOM K1 obs in C2;
STOR> INTEGER 1 6.
STOR> LET C3(K2)=MEAN(C2)
STOR> LET K2=K2+1
STOR> ECHO
STOR> END
MTB > NOTE   Assign values to the arguments and run Macro
MTB > BASE=1000
MTB > LET K1=10                      # Sample size n=10
MTB > LET K2=1                       # Initialize row position
MTB > EXECUTE 'B:MACRO3' 40 times
MTB > DESCRIBE C3
 .  .  .
MTB > HISTOGRAM C3
 .  .  .
```

5.5.5 Sampling Distributions with Varying Sample Sizes

In this section the above macro is executed for samples sizes 2, 5, 10, 50, and 100 with 40 repetitions each. Sampling distributions of sample means are shown by the following histograms. As the sample size becomes larger, the distribution approaches a bell shaped normal distribution.

```
Histogram of C3
Sample size N = 2                  Sample size N = 5

Midpoint    Count                  Midpoint    Count
    1.5       5    *****               1.5       2    **
    2.0       3    ***                 2.0       2    **
    2.5       1    *                   2.5       4    ****
    3.0       4    ****                3.0       8    ********
    3.5       8    ********            3.5       4    ****
    4.0       5    *****               4.0      12    ************
    4.5       2    **                  4.5       5    *****
    5.0       6    ******              5.0       2    **
    5.5       4    ****                5.5       1    *
    6.0       2    **


Sample size N = 10                 Sample size N = 50

Midpoint    Count                  Midpoint    Count
    1.6       1    *                   2.8       1    *
    2.0       0                        3.0       1    *
    2.4       1    *                   3.2       4    ****
    2.8       6    ******              3.4      11    ***********
    3.2       8    ********            3.6      11    ***********
    3.6       8    ********            3.8       9    *********
    4.0      11    ***********         4.0       2    **
    4.4       4    ****                4.2       0
    4.8       1    *                   4.4       1    *


Sample size N = 100

Midpoint    Count
    3.2       1    *
    3.3       1    *
    3.4       7    *******
    3.5      12    ************
    3.6       9    *********
    3.7       6    ******
    3.8       3    ***
    3.9       0
    4.0       1    *
```

99

The following table summarizes the results of the simulation. The first column gives the sample size. The second and the third columns contain the mean and the standard deviation of the sampling distribution of the 40 sample means. The last column shows the theoretical standard deviation of the sampling distribution under the Central Limit Theorem. As the sample size increases, the mean of the sampling distribution of the sample means approaches the population mean, μ, which is 3.5000. The standard deviation of the sampling distribution should approach $\sigma/\sqrt{n}$ with larger n. The approximation is not very close according to this simulation experiment as seen in the table. It may be that 40 repetitions are not sufficient.

Sample size	Sampling distribution Mean	Stdev	$\sigma/\sqrt{n}$
2	3.725	1.358	1.32286
5	3.525	0.945	0.83665
10	3.510	0.6205	0.59160
50	3.554	0.2540	0.26450
100	3.5395	0.1533	0.18708

5.6 MORE ON COMMANDS

5.6.1 The Random Command

Syntax:

RANDOM	K observations into each column C...C
BERNOULLI	p = K
BINOMIAL	n = K, p = K
CHISQUARE	degrees of freedom = K
DISCRETE	values in C probabilities in C
F	df numerator = K, df denominator = K
INTEGER	discrete uniform distribution on K to K
NORMAL	[mean = K [standard deviation = K]]
POISSON	mean = K
T	degrees of freedom = K
UNIFORM	continuous distribution [on K to K]

The RANDOM command generates a random sample of K values from a distribution specified by the subcommand, and puts the results in each column listed. Without any subcommand, the NORMAL distribution with mean=0 and standard deviation=1 is assumed.

Without the BASE command which sets the starting point for the random number generator, RANDOM produces different numbers each time you execute it.

Subcommands require arguments as specified above. For example, for BERNOULLI trials resulting in success(1) or failure(0), the probability of success (p) must be specified.

The BINOMIAL subcommand produces the number of successes after n Bernoulli trials with p probability of success on each trial.

The INTEGER subcommand produces integer values in the interval from K to K, each occurring with equal probability.

5.6.2 The BASE Command

Syntax:

```
BASE      K
```

The BASE command fixes a beginning value for the random number generator invoked by the RANDOM command. Without this command, RANDOM produces different numbers each time it is executed.

5.6.3 The SAMPLE Command

Syntax:

```
SAMPLE  K rows from columns C...C, put in C...C
        REPLACE
```

The SAMPLE command allows you to sample K rows at random from a block of columns. The sampled rows are stored in the columns specified. Without the subcommand REPLACE, sampling is done without replacement. The subcommand REPLACE generates a sample with replacement.

5.6.4 The MACRO Command

Syntax:

```
STORE   the following commands in the file 'filename'
(Minitab commands and subcommands)
(with optional arguments K)
END of stored commands

[LET K = value]
EXECUTE the commands in 'filename' [K number of times]
[NOECHO]
```

Using Macros involves the following steps:

1. Store a set of Minitab commands in a file: Begin with the STORE command together with the name of a file which will contain the commands. After STORE, type commands and subcommands which you wish to execute repetitively. This part may contain an argument or arguments whose values will be specified at the time of execution. At the end of the stored commands, type the END command.

2. Execute a set of commands by the EXECUTE command followed by the name of the file. If an unspecified argument is contained in the file, you must assign it a value before executing the macro file.

3. In any column number, the integer part of the column number may be replaced by a stored constant.

```
STORE 'MACRO9'
PRINT CK1
END
LET K1=5
EXECUTE 'MACRO9'          # This will PRINT C5
```

4. You can loop through a set of stored commands as many times as is specified by K after the EXECUTE command.

5. Without the NOECHO command, stored commands will be displayed on the screen while the macro file is executed. After suppressing the display by the NOECHO command, you can turn the echo on by the ECHO command.

```
NEW MINITAB COMMANDS

    RANDOM;
       BERNOULLI, BINOMIAL, INTEGER, DISCRETE, NORMAL
    PDF, CDF, INVCDF;
       BINOMIAL, NORMAL, INTEGER, POISSON
    SAMPLE
    STORE, END, EXECUTE, ECHO, NOECHO, BASE
    INDICATOR
```

EXERCISES

1. Compute the mean and standard deviation of GPA for the entire class.

2. From the class survey, sample 3 cases, and compute the mean GPA for the sample.

3. Take 10 samples of size 3. Compute the mean GPA in each sample.

4. Compute the means and standard deviations of the ten sample means. Compare them to the population mean and standard deviation.

5. Repeat Steps 3 and 4 by executing a macro 50 times.

CHAPTER SIX
TESTS OF MEANS

```
In This Chapter You Will Learn To:

    1.  Make point and interval estimations for a one sample
        mean
            With sigma known and unknown
    2.  Make point and interval estimations for two sample
        means
            With two independent samples and
            With two dependent samples
```

Having studied Minitab techniques for analyzing probability and sampling distributions, you are now ready to proceed to inferential statistics. The remaining five chapters of this book are concerned with making estimations about the characteristics of a population based on sample data. We will leave the class survey data at this point, and work with various types of research data which require appropriate statistical treatments. Within each chapter research problems are posed, then Minitab solutions are provided.

From among the various tasks in inferential statistics, the testing and estimation of means are selected for discussion in this chapter. Minitab addresses the testing of means, but does not provide pre-packaged programs for other problems such as estimating confidence intervals for proportions or variances.

6.1 ONE SAMPLE MEAN

6.1.1 Confidence Interval: Sigma Known

Problem

In response to the criticism that American education is deteriorating, a state university in recent years has raised entrance requirements. They no longer accept high school GPA's at face value but give their own minimum qualification test to incoming freshmen. From test results of the past several years, the mean is known to be 70 with a standard deviation of 9.2. Last year high schools finally began tightening their requirements also. Therefore this year the school administration expects the quality of students to improve. A professor, uncertain of such optimism on the administration's part, takes a random sample of 30 incoming freshmen, and examines their test scores. The scores are shown in Exhibit 6.1.

Based on this sample, can we conclude that student quality is improving? Construct a 95% confidence interval for the true mean if the sample mean is used as an estimate of the true average score of the qualifying test.

Exhibit 6.1

```
71 89 65 55 76 64 82 53 75 76
83 74 75 64 69 66 67 45 56 76
69 68 70 73 76 65 66 56 76 69
```

Minitab Solution : The ZINTERVAL Command

There is a Minitab command, ZINTERVAL, which is used to obtain a confidence interval for the mean when sigma is known.

Syntax:
```
ZINTERVAL   [K% confidence] assumed sigma=K,
            for data in C...C
```

The ZINTERVAL command calculates a confidence interval for the mean of variable(s) with standard deviation σ = K. This interval ranges from $\bar{X}-Z(\sigma/\sqrt{n})$ to $\bar{X}+Z(\sigma/\sqrt{n})$, where Xbar is the mean of the sample, n is the sample size, and Z is the value from the normal distribution corresponding to K percent confidence. If confidence level is not specified, the default level of 95% is used.

```
MTB > SET C1
DATA> 71 89 65 55 76 64 82 53 75 76
  . . .
DATA> END
MTB > ZINTERVAL 99% confidence assumed sigma=9.2, for C1

  THE ASSUMED SIGMA =9.20


            N       MEAN    STDEV  SE MEAN   99.0 PERCENT C.I.
  C1       30      68.97     9.47     1.68  (   64.63,    73.30)
```

The Minitab output reveals that the mean score in this sample of 30 students is 68.97, which is lower than the population mean of the past, 70. The sample standard deviation is 9.47. SE MEAN stands for the standard error of the mean, $\sigma/\sqrt{n}$. It equals $9.2/\sqrt{30}$ which computes to 1.68. The upper boundary of the 99% confidence interval is obtained as follows:

$$\text{MEAN} + Z*(\text{SE MEAN}) = 68.97 + 2.58*1.68 = 73.30$$

The lower boundary of the interval can be obtained similarly. You can be 99% confident that the interval [64.63, 73.30] will enclose the true population mean.

6.1.2 Hypothesis Testing: Sigma Known

Problem

Instead of estimating the mean of the population, the skeptical professor may test a two-tailed hypothesis, expecting no difference between this year's and previous years' students.

Minitab Solution: The ZTEST Command

To test a hypothesis concerning a sample mean with sigma known, Minitab offers the ZTEST command.

Syntax:
```
ZTEST   [of mu=K]assumed sigma = K on C...C
        ALTERNATIVE = K
```

The ZTEST command performs a Z test on one or more columns. Without the subcommand, it computes the appropriate p value based on a two-tailed test.

The subcommand ALTERNATIVE is used to perform a one-tailed test as follows:

```
Alternative = -1    Test the alternative hypothesis ( mu < K )
Alternative = +1    Test the alternative hypothesis ( mu > K )
```

For a two-tailed test, set ALTERNATIVE = 0 or run ZTEST without a subcommand.

Let us test the two-tailed hypothesis by the professor. The null hypothesis is set as: mu=70

```
MTB > # Test of Hypothesis
MTB > ZTEST mu=70 sigma=9.2 data in C1

  TEST OF MU = 70.000 VS MU N.E.  70.000
  THE ASSUMED SIGMA = 9.20

            N       MEAN    STDEV    SE MEAN         Z    P VALUE
  C1       30     68.967    9.467      1.680     -0.62       0.54
```

106

With a p value of 0.54 we fail to reject the null hypothesis that the true average score of this year is 70 (mu=70). The professor cannot say that this year's students are different from the previous ones.

Unbeknownst to the professor, a pessimistic colleague of his entertains the hypothesis that student quality has actually declined regardless of what school or college administrations have to say. He, therefore, takes the data collected by the professor to test this notion.

In this case, the colleague is exploring the one-tailed alternative hypothesis that mu<K. Accordingly, he uses the ALTERNATIVE subcommand with -1 in ZTEST. The null hypothesis remains the same, i.e., mu=70. For this one-tailed test, if the null hypothesis is rejected, then the claim that student quality has improved or remains the same (mu>=70) is disconfirmed at the same time.

```
MTB > ZTEST mu=70 sigma=9.2 data in C1;
SUBC> ALTERNATIVE=-1.


 TEST OF MU = 70.000 VS MU L.T. 70.000
 THE ASSUMED SIGMA = 9.20

          N      MEAN     STDEV    SE MEAN        Z    P VALUE
 C1      30     68.967     9.467     1.680     -0.62      0.73
```

This output with a p value of 0.73 fails to reject the null hypothesis that the true mean score is equal to 70. The colleague's pessimism regarding the decline in student quality may not be warranted.

6.1.3 Confidence Interval: Sigma Unknown

Problem

A newspaper reports that the average mortgage rate this month is 10.24% nationwide. A consumer group wishes to compare the mortgage rates available in their community. They took a random sample of mortgage companies, and gathered information as shown in Exhibit 6.2.

Construct a 95% confidence interval which includes the true mean mortgage rate if the sample mean is used to estimate the true mean.

Exhibit 6.2

```
10.25    10.25    10.125    9.99     9.875
10.00    10.00     9.75    10.00    10.00
 9.99    10.125   10.125    9.875   10.375
10.25    10.00    10.00    10.125    9.875
10.00    10.25    10.00    10.25    10.00
 9.99     9.875    9.875   10.125   10.375
10.25    10.125   10.25    10.25    10.70
10.125   10.25    10.125   10.50    10.2
10.25    10.375
```

Minitab Solution: The TINTERVAL Command

When the standard deviation of the population is not known, Minitab provides the TINTERVAL command to calculate a confidence interval for the mean. The default confidence level is 95%.

Syntax:
```
TINTERVAL    [K% confidence] for data in C...C
```

```
MTB > SET C2
DATA> 10.25  10.25 ...
DATA> END
MTB > TINTERVAL for data in C2

          N      MEAN     STDEV    SE MEAN    95.0 PERCENT C.I.
   C2    42    10.1219    0.1893    0.0292    ( 10.0629, 10.1809)
```

The output shows that you can be 95% sure that the interval between 10.06% and 10.18% contains the true mean mortgage rate, given the sample mean of 10.12% based on 42 lenders. The national average, 10.24%, reported in the paper is outside the interval. It means that a true rate as high as 10.24% is very unlikely for this particular community.

6.1.4 Hypothesis Testing: Sigma Unknown

Problem

If we are interested in testing the hypothesis that the true mean mortgage rate is 10.24% in this community, a two-tailed t test is appropriate. The null hypothesis states that the true mean is 10.24 against the alternative hypothesis that the true mean is different from 10.24.

Minitab Solution: The TTEST Command

To perform a t test, Minitab has the TTEST command.

Syntax:
```
TTEST    [mu=K] on C...C
ALTERNATIVE = K
```

The TTEST performs a two-tailed t test for the sample mean of each column listed where the population standard deviation is not known. It computes the t value, and displays the probability (p) associated with the computed t from a t distribution with (n-1) degrees of freedom, and a mu = K.

The subcommand ALTERNATIVE is used to specify the direction in a one-tailed test. See the explanation for the ZTEST command above.

```
MTB > TTEST mu=10.24 on C2

TEST OF MU = 10.2400 VS MU N.E. 10.2400

            N       MEAN      STDEV     SE MEAN          T     P VALUE
C2         42    10.1219     0.1893      0.0292      -4.04      0.0002
```

With a small p value of 0.0002, we reject the null hypothesis that the true average mortgage rate is 10.24% in this community. It agrees with the earlier finding that 10.24 lies outside the 95% confidence interval constructed around the sample mean.

6.1.5 Simulation of Confidence Intervals

Problem

To understand the concept of confidence interval better, let us simulate the construction of many intervals. If we construct 90% confidence intervals using 20 samples, two of the 20 intervals should fail to contain the true mean.

Exhibit 9.2 in Chapter 9 shows violent crime rates per 100,000 resident population in 50 states and the District of Columbia in 1988. Take a random sample of size 10, and construct a 90% confidence interval. Repeat this process 20 times and examine whether or not all the confidence intervals contain the true average murder rate.

Minitab Solution

First, examine the characteristics of the population, i.e., violent crime rates in 50 states (and DC) statistically and graphically.

```
MTB > SET C1
DATA> 157 148  ...  257
DATA> END
MTB > # Violent crime rates in 50 states (and DC), 1988
MTB > # per 100,000 population
MTB > NAME C1 'Crime'
MTB > DESCRIBE C1


            N      MEAN    MEDIAN    TRMEAN     STDEV    SEMEAN
Crime      51     494.9     452.0     462.4     320.5      44.9

          MIN       MAX        Q1        Q3
Crime    59.0    1922.0     273.0     653.0


MTB > HISTOGRAM 'Crime'


 Histogram of Crime    N = 51

 Midpoint    Count
        0        1   *
      200       14   **************
      400       14   **************
      600       12   ************
      800        6   ******
     1000        2   **
     1200        1   *
     1400        0
     1600        0
     1800        0
     2000        1   *
```

The mean crime rate is 494.9 per 100,000 population, and the standard deviation is 320.5. The histogram shows that the distribution is not normal.

To take 20 samples we create a macro, and execute it 20 times. Within the macro file MAC6.MTB in drive B, the SAMPLE command draws a random sample of 10 from C1 (from the previous SET C1), and stores the sample in column CK1. The column position CK1 is defined by the changing value of K1 as K1 is incremented by one, each time the macro is run. Before the macro is executed, K1 is set to 11. Therefore the first sample is stored in C11; the second sample will be in C12; and so on.

After a sample of 10 states is drawn, a 90% confidence interval is computed based on the sample data. Increment K1 by one after TINTERVAL is run.

```
MTB > STORE 'B:MAC6'
STOR> NOECHO
STOR> SAMPLE 10 from C1 put into CK1    #CK1 stores col pos
STOR> TINTERVAL 90 confidence for sample in CK1
STOR> LET K1=K1+1
STOR> ECHO
STOR> END
MTB > LET K1=11
MTB > BASE=1000
MTB > EXECUTE 'B:MAC6' 20 times
```

	N	MEAN	STDEV	SE MEAN	90.0 PERCENT C.I.	
C11	10	431.0	167.0	52.8	(334.2,	527.8)
C12	10	566.5	569.1	180.0	(236.5,	896.5)
C13	10	421.9	204.7	64.7	(303.2,	540.6)
C14	10	482.4	202.7	64.1	(364.8,	600.0)
C15	10	511.3	529.4	167.4	(204.3,	818.3)
C16	10	611.8	508.9	160.9	(316.7,	906.9)
C17	10	304.6	166.6	52.7	(208.0,	401.2)
C18	10	519.8	342.3	108.2	(321.3,	718.2)
C19	10	555.8	253.1	80.0	(409.0,	702.6)
C20	10	388.3	313.2	99.0	(206.7,	569.9)
C21	10	471.5	213.0	67.3	(348.0,	595.0)
C22	10	412.5	278.6	88.1	(251.0,	574.0)
C23	10	486.3	261.8	82.8	(334.5,	638.1)
C24	10	479.5	315.2	99.7	(296.7,	662.3)
C25	10	584.9	526.9	166.6	(279.4,	890.4)
C26	10	532.0	209.1	66.1	(410.8,	653.2)
C27	10	592.8	301.2	95.2	(418.2,	767.4)
C28	10	498.4	197.1	62.3	(384.1,	612.7)
C29	10	529.0	212.7	67.3	(405.7,	652.3)
C30	10	376.2	156.7	49.6	(285.3,	467.1)

Upon inspection, samples stored in C17 and C30 produced confidence intervals [208.0-401.2] and [285.3-467.1] that do not contain the population mean 494.9. Two out of 20, or 10% failure agrees with the 90% confidence principle.

6.2 TWO SAMPLE MEANS

6.2.1 Two Independent Samples

Problem

An auto safety group ranked tires based on the U.S. testing program, which gives credit to special traction and heat resistance. Based on this test, they have classified longest lasting tires into two groups: normal and super. A consumer group wished to find out if super tires, that is tires with special performance features would also get mileage comparable to standard tires. Mileage of tires is measured by the tread-wear ratings of tires multiplied by 200 under test conditions including special temperature. They took random samples of tires from brands classified as normal and super to compare their mileages. The results are shown in Exhibit 6.3. Is there a significant difference between two groups of brands in their mileage performance?

Exhibit 6.3

Normal	Super
80000	68000
80000	68000
80000	68000
74000	68000
74000	68000
74000	68000
74000	64000
74000	58000
74000	58000
74000	56000
	56000
	54000
	54000
	54000
	54000

Minitab Solution: The TWOSAMPLE Command

Minitab provides the TWOSAMPLE command to perform a t test of the difference in two sample means. The two samples are assumed to be independent. This command calculates a confidence interval as well as a t statistic and its associated p value.

Syntax:
```
TWOSAMPLE  [K% confidence] for data in C and C
         ALTERNATIVE = K
         POOLED
```

TWOSAMPLE specifies two sample data in two columns. The first C contains a sample from population 1, with mean μ_1, while the second C contains a sample from population 2, with mean μ_2. TWOSAMPLE without subcommands performs a two-tailed t test of Ho:$(\mu_1=\mu_2)$ and constructs a confidence interval around $(\mu_1-\mu_2)$. If the confidence

112

level is not specified, 95 percent is assumed. If the subcommand POOLED is not used, TWOSAMPLE uses a procedure that does not assume equal variances of the two populations.

If the subcommand ALTERNATIVE is used, it performs a one-tailed test. The use of ALTERNATIVE is the same as described for the ZTEST command.

Before running a t test, we need to test the equality of variances to determine whether the subcommand POOLED should be used. Minitab does not provide a direct test, but an F ratio can be calculated by using Minitab functions and arithmetic operations.

Obtain the standard deviations (STDEV) of two samples, and square them to compute the variances. An F ratio is the ratio of these two variances with appropriate degrees of freedom. For the degrees of freedom, first use COUNT to determine the sample size, then subtract 1 from it.

```
MTB > # Equal variance test
MTB > STDEV C1 K1
      ST.DEV. =        2898.3      # standard deviation  Normal
MTB > STDEV C2 K2
      ST.DEV. =        6363.6      # standard deviation  Super
MTB > COUNT C1 K3
      COUNT   =     10             # sample size   Normal
MTB > COUNT C2 K4
      COUNT   =     15             # sample size   Super
MTB > LET K5=K1**2                 # variance   Normal
MTB > LET K6=K2**2                 # variance   Super
MTB > LET K8=K6/K5                 # F-ratio   (Super/Normal)
MTB > PRINT K8
 K8         4.82086
MTB > #                   (Super)df= 15-1=14      (Normal)df=10-1=9

MTB > INVCDF .95;                  # alpha=.05
SUBC> F 14 9.
      0.9500      3.0255
```

The F value for 'Normal' and 'Super' tire samples is 4.82086. Using the 0.05 level of significance, INVCDF for the F distribution with degrees of freedom 14 and 9 produces the critical F value of 3.0255. Since 4.82086 is larger than 3.0255, we reject the null hypothesis of equal variance at the 0.05 level.

TWOSAMPLE without the subcommand POOLED performs a t test using separate estimations of unequal variances.

The output shows that the mean mileage for ordinary tires is 75,800 miles, while the mean for super performance tires is 61,067 miles. The standard deviations are 2,898 and

6,364 miles respectively. The two variances which are the squares of these standard deviations seem to be quite different, which justifies using a t test based on separate estimates of variances.

```
MTB > NAME C1 'Normal' C2 'Super'
MTB > TWOSAMPLE t for data in  C1 and C2

 TWOSAMPLE T FOR Normal VS Super
            N       MEAN      STDEV     SE MEAN
 Normal    10      75800       2898         917
 Super     15      61067       6364        1643

 95 PCT CI FOR MU Normal - MU Super: (10808, 18659)

 TTEST MU Normal=MU Super (VS NE): T= 7.83   P=0.0000   DF=   20
```

A 95% confidence interval for the difference between the true means of the two kinds of tires ('Normal'-'Super') is [10,808 - 18,659], which is estimated based on the difference between sample means (75,800-61,067=14,737). With t=7.83 and p=0.0000 (=0.0005), the null hypothesis that the mean mileages of 'Normal' and 'Super' tires are the same is rejected. There is indeed a difference between the two kinds of tires in their mileages, and the difference is not necessarily in the direction as the consumers would like to see.

If an F test shows that variances are equal, use the POOLED subcommand under the TWOSAMPLE command. With this subcommand the common variance of the two populations is estimated by the pooled sample variance. The test is based on the statistic $T = (\overline{x}_1 - \overline{x}_2)/s$ with (n1+n2-2) degrees of freedom. To compare the results, let us run a t test with POOLED for the same set of data.

```
MTB > TWOSAMPLE T C1 C2;
SUBC> POOLED.

 TWOSAMPLE T FOR Normal vs Super
            N       MEAN      STDEV     SE MEAN
 Normal    10      75800       2898         917
 Super     15      61067       6364        1643

 95 PCT CI FOR MU Normal - MU Super: (10269, 19198)

 TTEST MU Normal=MU Super (VS NE): T= 6.83   P=0.0000   DF=   23

 POOLED STDEV =         5285
```

With a p value of 0.0000(=0.00005) we again reject the null hypothesis that the two types of tires have the same mileage performance. Note that the confidence interval is wider, and the t value is smaller under this equal variance model as compared to the unequal variance model. Using a common variance model when variances are in fact very different is therefore inefficient.

6.2.2 Two Independent Samples Stacked in One Column

Problem

Sometimes data are arranged in such a way that the values of the two samples are in one column. You may wish to compare these values according to a classificatory variable stored in another column. For example, the mileages for all tires may appear in C3, and the 'Normal' vs 'Super' distinction may appear in C4.

Minitab Solution: The TWOT Command

For this type of data arrangement, Minitab offers another t test command, called, TWOT.

Syntax:

```
TWOT  [K percent confidence] data in C, groups in C
      ALTERNATIVE = K
      POOLED
```

The TWOT command performs the same test as TWOSAMPLE. The only difference is in the organization of the sample data. TWOSAMPLE expects the first sample data to be located in the first column, and the second sample data in the second column. In contrast, TWOT assumes that the data for both groups are in the first column where they may be mixed, while the second column specifies to which group each observation belongs.

Again for comparison, let us stack the data on 'Normal'(C1) and 'Super'(C2) tires into one column(C3) by the STACK command. The SUBSCRIPTS subcommand stores the group identification in C4: 1 for Normal and 2 for Super. Thus TWOT can be run on data in C3 and group in C4 as follows:


```
MTB > STACK C1 C2 C3;
SUBC> SUBSCRIPTS C4.
MTB > PRINT C1-C4
```

ROW	Normal	Super	C3	C4
1	80000	68000	80000	1
2	80000	68000	80000	1
3	80000	68000	80000	1
4	74000	68000	74000	1
5	74000	68000	74000	1
6	74000	68000	74000	1
7	74000	64000	74000	1
8	74000	58000	74000	1
9	74000	58000	74000	1
10	74000	56000	74000	1
11		56000	68000	2
12		54000	68000	2
13		54000	68000	2
14		54000	68000	2
15		54000	68000	2
16			68000	2
17			64000	2
18			58000	2
19			58000	2
20			56000	2
21			56000	2
22			54000	2
23			54000	2
24			54000	2
25			54000	2

```
MTB > TWOT data in C3 groups in C4

TWOSAMPLE T FOR C3
C4    N       MEAN      STDEV    SE MEAN
1    10      75800      2898        917
2    15      61067      6364       1643

95 PCT CI FOR MU 1 - MU 2: (10808, 18659)

TTEST MU 1 = MU 2 (VS NE): T= 7.83   P=0.0000   DF=   20
```

As you see, the results are identical to those produced by TWOSAMPLE in the earlier section.

6.2.3 Two Dependent Samples

Problem

With the onset of the Persian Gulf crisis gasoline prices jumped in one week, although some gas stations were still trying to maintain prices at the level before the crisis.

Ever in search of cheaper gas and customer-friendly gas stations, a consumer took a random sample of 12 gas stations in his community, and made a price comparison of one week before and after the Iraqi occupation of Kuwait.

This situation calls for a comparison of means of two dependent samples, since the same gas stations constitute the before and after samples. A t test for paired samples is required to test the null hypothesis that there is no difference in gasoline prices before and after the Gulf crisis.

Exhibit 6.4

1	.89	1.24
2	.86	1.19
3	.88	1.24
4	.89	.89
5	.76	.83
6	.97	1.33
7	.88	.88
8	.91	1.34
9	.87	1.11
10	.89	1.25
11	.87	1.23
12	.90	.90

Minitab Solution

There is no specific Minitab command to run a t test for paired samples. Compute the difference between the two dependent samples case by case. Then run TINTERVAL and TTEST for the difference scores stored in C3. The paired sample t-test is equivalent to the one sample t-test with a mu = 0. In the following TTEST command, a mu value is not specified after the command keyword. When a value is not given for mu, Minitab computes the test according to the null hypothesis of mu = 0.

```
MTB > # Gas stations
MTB > READ C1-C2
MTB > NAME C1 'Before'  C2 'After'
MTB > LET C3=C2-C1
MTB > NAME C3 'Diff'
MTB > TINTERVAL 99% confidence using the difference in C3
```

	N	MEAN	STDEV	SE MEAN	99.0 PERCENT C.I.
Diff	12	0.2383	0.1694	0.0489	(0.0864, 0.3902)

```
MTB > TTEST C3

TEST OF MU = 0.0000 VS MU N.E.  0.0000

            N      MEAN     STDEV    SE MEAN         T     P VALUE
Diff       12    0.2383    0.1694     0.0489      4.87      0.0005
```

The mean of the differences between before and after gas prices is 23.83 cents per gallon. Based on this data consumers can be 99% confident that the true mean price hike is between 8.64 and 39.02 cents per gallon.

The null hypothesis that there is no price difference is rejected with a p value as small as 0.0005.

NEW MINITAB COMMANDS

   ZINTERVAL, TINTERVAL
   ZTEST, TTEST
   TWOSAMPLE, TWOT

EXERCISES

1. Test the hypothesis that the average grade expected for the class is C or 2.

2. Test the hypothesis that freshmen and seniors spent the same time studying.

3. Test the hypothesis that satisfaction with college remains the same.

CHAPTER SEVEN
ANALYSIS OF VARIANCE

```
In This Chapter You Will Learn To:

    1.  Do a oneway analysis of variance
    2.  Make multiple comparisons of pairwise means
    3.  Do an analysis of variance with randomized block
          design
    4.  Do twoway analysis of variance with balanced and
          unbalanced data
```

In the previous chapter, the hypothesis was tested that two population means are equal by comparing the two sample means derived from these populations. This chapter deals with hypothesis testing which involves three or more populations. The test which examines the equality of means among two or more groups or samples is the analysis of variance. The groups in an analysis of variance are defined by one or more independent variables called factors.

7.1 ONEWAY ANALYSIS OF VARIANCE

7.1.1 AOVONEWAY Analysis of Variance

<u>Problem</u>

In recent years the Japanese style management has been considered an efficient alternative to traditional American management. Instead of the individualistic contractual relationship hitherto prevalent in American businesses and industries, a paternalistic collectivistic orientation of the Japanese corporate culture is being introduced.

To test the hypothesis that the Japanese style of labor relationship is likely to decrease absenteeism, an experiment was undertaken in a large manufacturing company. Twelve work groups were formed and randomly assigned to three types of work relations: (1) bureaucratic, (2) paternalistic, and (3) contractual. These types of relations were created experimentally by the foremen as well as by the laborers themselves. The resulting absenteeism rates per employee after half a year of experimentation are presented in Exhibit 7.1. Test the null hypothesis that there is no difference in absenteeism among the different management styles.

Exhibit 7.1

Bureaucratic	Paternalistic	Contractual
6.3	4.4	5.1
7.1	5.2	5.5
4.9	3.6	5.6
6.8	5.4	3.5

Minitab Solution: The AOVONEWAY Command

The AOVONEWAY command performs a oneway analysis of variance, assuming the observations for each level or group are in different columns. Minitab refers to a group or level as a cell. There must be two or more cells for AOVONEWAY analysis, although an equal number of observations in each cell is not necessary.

Syntax:

```
AOVONEWAY   on data in C,...,C
```

The three types of management are entered in C1-C3. After AOVONEWAY, specify those columns containing absenteeism under the three levels of management.

```
MTB > #  Oneway   AOVONEWAY
MTB > #  Absenteeism and Management style
MTB > READ C1-C3
DATA> 6.3  4.4   5.1
DATA> 7.1  5.2   5.5
DATA> 4.9  3.6   5.6
DATA> 6.8  5.4   3.5
DATA> END
MTB > NAME C1 'Bureauc' C2 'Paternal'  C3 'Contract'
MTB > AOVONEWAY C1-C3
```

ANALYSIS OF VARIANCE

SOURCE	DF	SS	MS	F	p
FACTOR	2	6.052	3.026	3.53	0.074
ERROR	9	7.725	0.858		
TOTAL	11	13.777			

```
                                   INDIVIDUAL 95 PCT CI'S FOR MEAN
                                   BASED ON POOLED STDEV
LEVEL      N     MEAN     STDEV   ----------+---------+---------+----
Bureauc    4   6.2750    0.9743                 (-------*--------)
Paternal   4   4.6500    0.8226   (--------*-------)
Contract   4   4.9250    0.9743    (--------*--------)
                                  ----------+---------+---------+----
POOLED STDEV =    0.9265                    4.8       6.0       7.2
```

120

The total sum of squares is broken down into its two sources: the variation due to differences among the three management types (SS Factor) and the variation due to random error within a management style (SS Error). Thus (SS Total)=(SS Factor)+(SS Error).

The fourth column in the analysis of variance table gives the mean square due to the factor (differences among management types) and the mean square due to error. Each mean square is the ratio of the corresponding sum of squares and its degrees of freedom (DF). The fifth column gives the quotient of these two mean squares:
(F ratio)=(MS Factor)/(MS Error).

If this F ratio is large, then the MS Factor must be much larger than the MS Error. In other words, the variation among the management types is greater than the variation due to random error. The p value of 0.074 in the last column gives the probability of the F ratio of 3.53 or larger under the null hypothesis of no difference in absenteeism among three management types. If we had used a 0.10 or 0.075 level of significance, we would have rejected the null hypothesis. However, using a 0.05 level of significance, we fail to do so. The overall difference in absenteeism under the three management styles does not seem to be significant at the 0.05 level.

Below the analysis of variance table, the output displays for each level: the number of cases, mean, standard deviation, and individual 95% confidence interval for the mean based on pooled standard deviations.

7.1.2 ONEWAY Analysis of Variance

Problem

Frequently, data are entered all in one column with their subgroup identifications stored in another column. For such an arrangement of data, Minitab provides another command, ONEWAY, to perform oneway analysis of variance.

Minitab Solution: The ONEWAY Command

Syntax:
```
ONEWAY analysis of variance, data C, subscripts C
        [put residuals in C [fits into C]]
```

The ONEWAY command requires that all data are in one column, and that a second column gives the subscripts which are the level, cell or group identifications. The subscripts must be integers between -10,000 and +10,000 or missing values.

The optional third column stores residuals (observed value minus cell means). If a fourth column is specified, it stores the fitted values (cell means).

Let us transform the previous data into a format amenable to ONEWAY. The STACK command stacks C1 on top of C2, which is stacked on top of C3. The stacked contents are then stored in an empty column C11. The SUBSCRIPTS subcommand keeps track of the group identification numbers in C13. To understand the stacking process better, you may want to print all the columns involved.

```
MTB > # Create columns by stacking
MTB > STACK C1 C2 C3 C11;
SUBC> SUBSCRIPTS C13.
MTB > ONEWAY data in C11, levels in C13
```

ANALYSIS OF VARIANCE ON C11

SOURCE	DF	SS	MS	F	p
C12	2	6.052	3.026	3.53	0.074
ERROR	9	7.725	0.858		
TOTAL	11	13.777			

```
                                    INDIVIDUAL 95 PCT CI'S FOR MEAN
                                    BASED ON POOLED STDEV
LEVEL      N      MEAN    STDEV  ---------+---------+---------+---
    1      4    6.2750   0.9743                 (-------*--------)
    2      4    4.6500   0.8226  (--------*-------)
    3      4    4.9250   0.9743    (--------*--------)
                                 ---------+---------+---------+---
POOLED STDEV =     0.9265            4.8       6.0       7.2
```

The results of ONEWAY are identical to those of AOVONEWAY. To visualize the distribution of absentee rates under the three management types, let us plot horizontal histograms.

```
MTB > DOTPLOT C11;
SUBC> BY C13.


C12
1                             .              .      .   .
   -+---------+---------+---------+---------+---------+-----C11

C12
2      .           .           .  .
   -+---------+---------+---------+---------+---------+-----C11

C13
3      .                  .         ..
   -+---------+---------+---------+---------+---------+-----C11
 3.50      4.20      4.90      5.60      6.30      7.00
```

7.1.3 Multiple Comparisons of Means

Problem

The F test tells us whether or not the hypothesis about the equality of means among several groups should be rejected. However, it does not specify which particular group mean or group means are significantly different from one another. Accordingly, following a significant F test, the next step in the inquiry is to determine which specific pair or pairs of comparison, or whether all pairs, are causing the difference.

On the other hand, the non-rejection of a null hypothesis does not end the analysis either. We may still be interested in knowing if there is a significant difference in a particular pairwise comparison such as between the paternalistic and bureaucratic, or between the paternalistic and contractual, although one significant difference may not be sufficient to raise the overall significance of the results. For example, an examination of the DOTPLOT as well as the output from analysis of variance indicates that the mean absenteeism rate in the bureaucratic work group is very different from the absenteeism rate found either in the paternalistic or contractual management style.

Minitab Solution

Four subcommands for multiple comparison are available under the ONEWAY command, namely, TUKEY, FISHER, DUNNETT, and MCB. TUKEY and FISHER make all pairwise comparisons among subgroups. DUNNETT makes a comparison of each treatment with user specified control groups. MCB, on the other hand, compares each group with the largest, or the smallest of the others. A desired alpha level can be specified under all four methods of comparison.

Let us use TUKEY to make all pairwise comparisons among the three management types. In particular, we are interested in knowing if the paternalistic management is significantly effective in reducing absenteeism as compared to the bureaucratic or contractual managements.

After the ONEWAY command, attach a subcommand TUKEY. You may specify the nominal or family error rate in parenthesis after TUKEY; otherwise, the default alpha value of 0.05 will be used. The output from the Tukey test shows three types of significance levels. First, there is the user-specified nominal or alpha level of significance. Since we have not specified a level, Minitab assumes the default value of 0.05. The family error rate, which is the same as the nominal or alpha level, refers to the maximum likelihood of one or more intervals being incorrect over all possible intervals calculated, while the individual error rate gives the probability that any particular interval is incorrect.

```
MTB > ONEWAY C11 C13;
SUBC> TUKEY.

 TUKEY'S multiple comparison procedure

            Nominal level  = 0.0500
      Family error rate    = 0.0500
 Individual error rate     = 0.0210

 Critical value = 3.95

 Intervals for (mean of column group) - (mean of row group)

                1              2

 2      -0.2028
         3.4528

 3      -0.4778        -2.1028
         3.1778         1.5528
```

The critical value, 3.95, can be found in the Studentized Range Distributions Table with k and v degrees of freedom where k refers to the number of levels or groups, and v equals N - k, with N = the total number of subjects. The probability of the critical value is determined by $P(cv)=1$-alpha, where alpha is the family error rate. In this case, $P(3.95)=0.95$. This critical value is used (in much the same way as the z or t scores) for the calculation of the confidence intervals which follow in the display.

Thus at the bottom of the output are pairs of numbers indicating the lower and upper limits of the confidence intervals for the difference of the means between the column and the row group. For example, the confidence interval for the difference of means of group 1 (the column group) and group 2 (the row group) is -.2028 to 3.4528. An inspection of the three confidence intervals shows that all three contain "0" within their interval. This means that it is possible for the true difference between the two means to be zero. Recall that when the null hypothesis value is contained within the confidence interval, you cannot reject the null hypothesis. In short, we fail to reject the null hypothesis of no difference between groups 1 and 2, groups 1 and 3, and groups 2 and 3. None of the pairwise comparisons among the three management styles shows significant difference.

7.2 RANDOMIZED BLOCK DESIGN

Problem

The randomized block design uses blocks of subjects as if these blocks constitute a second independent variable. As an extension of the paired difference design, it makes comparisons among different treatments within blocks of relatively homogeneous experimental subjects.

An investor is faced with the decision of selecting a financial consultant who will manage his $100,000 investment. Three managers have been introduced to him. Their orientations are all rated as medium to aggressive with goals aimed at growth rather than income. To help him decide, he compares the average annual returns of their managed portfolios during the past nine years. The results are shown in Exhibit 7.2. Is there a significant difference among the three consultants?

Exhibit 7.2

Year	Consultants		
	A	B	C
1981	6.08	10.42	4.74
1982	32.30	30.01	29.92
1983	28.17	29.23	17.87
1984	10.69	9.17	11.63
1985	30.36	24.70	37.71
1986	15.27	17.29	19.38
1987	2.59	2.29	8.04
1988	9.72	21.83	12.72
1989	20.86	22.80	24.55

Minitab Solution: The TWOWAY Command

The TWOWAY command performs twoway analysis of variance for balanced data (equal number of observations, one or more, in each cell). TWOWAY is appropriate for the randomized block design because an additive model (a model without an interaction term) is fitted if there is one observation per cell or if the subcommand ADDITIVE is used. In the above experiment, each year is considered "a block of subjects". The treatment refers to the management techniques of three financial consultants. In a randomized block design no interaction could exist conceptually between the treatments and blocks, as subjects are homogeneous within blocks. In practice, the treatment effects are assumed to be the same within each block of subjects.

Syntax:

```
TWOWAY analysis of variance, data in C,
       subscripts C, C
       [ store residuals in C [fits in C] ]

       ADDITIVE
       MEAN
```

Residuals and fitted values are optionally stored in the columns specified after the indication of data input and output in TWOWAY. The subcommand ADDITIVE adopts an additive model, that is, a model without the interaction term. In this case the fitted value for cell (i,j) is (row i mean) + (column j mean) - (grand mean). If the subcommand ADDITIVE is not used, the fitted value is the cell mean.

The subcommand MEANS produces marginal means for specified rows or columns, and their 95% confidence intervals.

In the following example, the data for TWOWAY are entered as follows. The first column contains annual average yields for each year for each financial consultant. The second column refers to the three consultants (1 to 3). The third column contains the values 1 to 9 to represent the nine years.

To enter C2 and C3, multipliers are used. The post multiplier for C2 produces the repetition of each element in the parentheses 9 times: 1 1 1 1 1 1 1 1 1 2 2 2 2 2 2 2 2 2 3 3 3 3 3 3 3 3 3. The pre-multiplier for C3 repeats the numbers 1 to 9 three times: 1 2 3 4 5 6 7 8 9 1 2 3 4 ... 1 2 3 ...

You can use the PRINT or TABLE command to confirm your data entry.

```
MTB > SET C1
DATA> 6.08   32.30   28.17   10.69   30.36   . . . . . . 24.55
DATA> END
MTB > SET C2
DATA> (1:3)9
DATA> END
MTB > SET C3
DATA> 3(1:9)
DATA> END
MTB > TWOWAY C1-C3;
SUBC> MEANS C2.
```

ANALYSIS OF VARIANCE   C1

SOURCE	DF	SS	MS
C2	2	9.2	4.6
C3	8	2327.8	291.0
ERROR	16	294.8	18.4
TOTAL	26	2631.7	

```
                                  Individual 95% CI
      C2      Mean    ---------+---------+---------+---------+--
       1      17.3    (--------------*--------------)
       2      18.6        (--------------*--------------)
       3      18.5       (--------------*--------------)
                      ---------+---------+---------+---------+--
                          16.0      18.0      20.0      22.0
```

126

```
MTB > LET K1=4.6/18.4          # F ratio for 3 consultants
MTB > LET K2=291/18.4          # F ratio for 9 years
MTB > PRINT K1 K2
 K1         0.250000
 K2         15.8152
MTB > INVCDF 0.95;
SUBC> F 2 16.                  # df=2 & 16
      0.9500      3.6338       # Critical value, 3 consultants
MTB > INVCDF 0.95;
SUBC> F 8 16.                  # df = 8 & 16
      0.9500      2.5911       # Critical value for 9 years
```

The analysis of variance table generated by TWOWAY, however, does not include the F statistics automatically. It is necessary to compute the F ratios by the LET commands, then compare them with the critical F values.

To test the null hypothesis that there is no difference among the three consultants, we calculate the ratio of MS C2 to MS Error. The resulting F statistic is 0.25. As you know, when the F value is less than 1, it is not significant. However, to double check the results, use INVCDF to compute the critical F value with 2 and 16 degrees of freedom, at the 0.05 level of significance which turns out to be 3.6338. Therefore the investor cannot reject the null hypothesis that the three consultants are the same.

On the other hand, although the investor is not primarily interested in the effects of yearly fluctuations, the F statistic could easily be computed. Thus, the F statistic for years is 15.8152, which is much larger than the critical F value of 2.5911 at the 0.05 significance level. The investor rejects the null hypothesis that yearly fluctuations are nil.

The MEAN subcommand produces the mean annual yields over nine years achieved by the three financial consultants together with their 95% confidence intervals. The overlap of the three confidence intervals confirms the result of the F test, that there is no significant difference in the yields among the management techniques of the three consultants.

7.3 TWOWAY ANALYSIS OF VARIANCE: BALANCED DATA

7.3.1 Two Factorial Analysis of Variance

Problem

Finally we are interested in determining simultaneously the effects of two variables and their interaction upon the third variable.

"Is it necessary to diet and also exercise if you want to lose weight?" Many overweight people ask this question because it is difficult enough to follow one routine, not to mention adhering to two regimens. A physical fitness institute conducted an experiment with 18 obese women in their 30s who weighed an average of 214 pounds.

The first factor is the diet which has three levels: (1) regular diet, (2) controlled diet of 1,200 calories a day, and (3) liquid diet (420 calories a day) for 16 weeks followed by the controlled diet.

The second factor is the amount of exercise with levels: (1) walking 10 miles a week, and (2) no special exercise.

The 12 women were randomly assigned to one of the 6 groups produced by these 2 factors. The weight losses or gains after 48 weeks are reported in Exhibit 7.3.

Exhibit 7.3

	Regular	Controlled	Liquid
Walking	20	-34	-45
	28	-36	-46
	24	-38	-40
No exercise	18	-20	-43
	17	-22	-42
	15	-23	-30

Does a special diet have a significant effect on weight loss? How about physical exercise? Is there any interaction between the two?

Minitab Solution: The TWOWAY Command

Without the subcommand ADDITIVE and with more than one observation per cell, TWOWAY performs a twoway analysis of variance of the full model, including the effect of interaction. In the solution below, C1 contains the observed weight gain or loss for 18 women. C2 refers to the two levels of physical activity: exercise walking(1) and no exercising(2). The SET command for C2 produces three repetitions of 1 and then 2, which in turn are repeated three times: 1 1 1 2 2 2 1 1 1 2 2 2 1 1 1 2 2 2. C3 contains the three types of diet: regular(1), controlled(2), and liquid(3).

TWOWAY specifies that the mean scores of 'Weight' are to be compared among subgroups produced by 'Walk' and 'Diet'.

```
MTB > SET C1
DATA> 20 28 24 18 17 15 -34 -36 -38 -20 ...
DATA> END
MTB > SET C2
DATA> 3(1:2)3
DATA> END
MTB > SET C3
DATA> (1:3)6
DATA> END
MTB > NAME C1 'Weight' C2 'Walk' C3 'Diet'
MTB > TWOWAY data in 'Weight' factors in 'Walk' and 'Diet';
SUBC> MEANS 'Walk' 'Diet'.

   ANALYSIS OF VARIANCE  Weight

   SOURCE          DF        SS         MS
   Walk             1       76.1       76.1
   Diet             2    12654.3     6327.2
   INTERACTION      2      355.4      177.7
   ERROR           12      174.7       14.6
   TOTAL           17    13260.5
```

Before examining the effect of an individual factor, we need to know whether or not the interaction effect is significant. The F statistic is calculated by dividing the interaction MS by the error MS, which results in 12.1712. INVCDF displays the critical F value at the 0.05 level with 2 and 12 degrees of freedom. Since the F statistic from this sample(12.1712) is larger than the critical value (3.8853), the null hypothesis of no interaction is rejected. We will postpone the discussion of the calculation of the F ratios for the two factors when significant interaction exists, since it is a situation which requires careful consideration. Instead, we will deal with the simpler case where there is no significant interaction effects first.

```
MTB > LET K1=177.7/14.6          # Interaction F value
MTB > PRINT K1
 K1        12.1712
MTB > INVCDF 0.95;
SUBC> F 2 12.
0.9500      3.8853
```

If interaction is not significant, we can include the interaction term in the error term. This is accomplished by the subcommand ADDITIVE in TWOWAY. Note in the output below that the MS's for 'Walk' and 'Diet' are the same as those obtained with the interaction term separated.

```
MTB > TWOWAY 'Weight' 'Walk' 'Diet';
SUBC> ADDITIVE.
```

```
ANALYSIS OF VARIANCE  Weight

    SOURCE          DF          SS          MS
    Walk            1         76.1        76.1
    Diet            2      12654.3      6327.2
    ERROR          14        530.1        37.9
    TOTAL          17      13260.5


MTB > LET K2=76.1/37.9          # F value for walk
MTB > PRINT K2
  K2        2.00792
MTB > INVCDF .95;
SUBC> F 1 14.
     0.9500      4.6001


MTB > LET K3=6327.2/37.9         # F value for diet
MTB > PRINT K3
  K3        166.945
MTB > INVCDF .95;
SUBC> F 2 14.
     0.9500      3.7389
```

The F statistic is calculated for each factor: diet and exercise, using the new error term. K2 represents the F ratio for 'Walk'. It equals 2.00792 and is less than the F critical value of 4.6001 at the 0.05 level of significance. Walking seems to have little effect upon weight loss.

As for the effect of diet, the F ratio as represented by K3 of 166.945 is larger than the critical F value of 3.7389 at the 0.05 level of significance. Therefore we reject the null hypothesis that diet has no effect on weight.

7.3.2 Dealing with Interaction

Problem

When interaction is present it is necessary to qualify the estimate of the effects of the independent variables. The effects of one independent variable should be interpreted for each category of the other independent variable. We should ask how a special diet may affect weight loss differently depending on whether or not exercise walking is involved.

LPLOT is frequently useful in examining the nature of the interaction. In the following scatterplot 'Weight' is plotted against 'Diet' simultaneously for two levels of physical exercises. The "A" symbol refers to walking, while "B" means no exercise. If there is no interaction, "A" should be stacked always above or below "B" under different diet treatments. In this plot "A" (walking) is above "B"(no exercise) under the regular diet ('Diet'=1), while it is below "B" under the controlled or liquid diet ('Diet'=2,3). If a person is not dieting at all, exercise seems to cause weight gain. On the other hand, when a person is on a special diet, then exercise helps to lose more weight.

```
MTB > LPLOT 'Weight' vs 'Diet' coded 'Walk';
SUBC> XINCREMENT=1;
SUBC> XSTART 0.
```

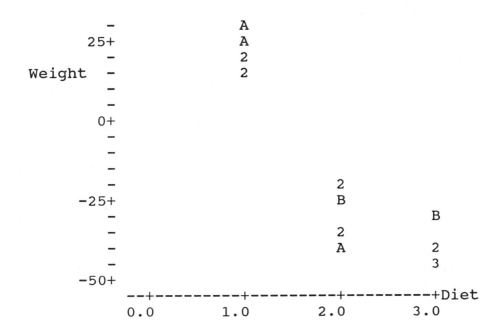

Once we have located where the interaction exists, what is the next step in the analysis? A decision must be made as to whether the error or the interaction term should be used as the denominator of the F statistic to determine the effects of independent variables. This decision depends on whether the categories of an independent variable are all inclusive or are sampled. The limitation of space does not allow an elaboration of this topic. However, one approach is suggested below.

In this experiment "exercise walking" versus "no exercise" are considered as dichotomous categories of "doing some" against "not at all". In view of the physical shape the subjects were in, walking was all they could take as routine exercise. On the other hand the three categories of 'Diet' are more likely to be considered as a sample from all possible dieting programs. In testing the effect of the nonsampled (fixed) factor, 'Walk', the effect of the sampling of the second factor must be considered. A second sample of different diets may produce a very different interaction estimate. Since diets were sampled for the experiment, it is possible that such interaction effects might accumulate in one or another of the categories of the fixed factor 'Walk'. Thus the sum of squares of 'Walk' may reflect not just the effect of 'Walk' but also the interaction effects. Therefore a conservative procedure would be to use the interaction estimate as the denominator of the F statistic.

In order to test a specific hypothesis including the interaction term, Minitab provides ANOVA with the TEST subcommand. Let us test the effect of 'Walk' using the interaction of 'Walk' and 'Diet' as the error term.

Minitab Solutions: The ANOVA Command

Syntax:

```
ANOVA        model

      RANDOM        factorlist
      EMS
      FITS          C,...,C
      RESIDUALS     C,...,C
      MEANS         termlist
      TEST          termlist /errorterm
      RESTRICT
```

The ANOVA command performs multiway analysis of variance for balanced designs, and oneway analysis of variance for unbalanced designs. Factors may be crossed or nested, fixed or random.

After the command keyword, ANOVA, specify a model with a dependent variable followed by an equal sign and independent variables. For two factors with interaction, a model is specified as either of the two below:

ANOVA Y=A B A*B
ANOVA Y=A|B

A*B denotes the interaction term between A and B. The symbol "|" signifies a full model created by two factors, which includes A, B, and A*B.

Only a few subcommands are explained here. For further information use the Help facility or the Minitab Reference Manual.

The subcommands, FITS, RESIDUALS, and MEANS, work in the same way as with TWOWAY.

The TEST subcommand is used to perform F tests which are not done automatically. As arguments for the subcommand, specify a list of one or more terms in the model for which you wish to construct F tests. Then enter a '/' as a delimiter. The error term which follows the slash is a term in the model to be used as the denominator in the F tests. Error term may also be a linear combination of terms from the model and ERROR representing MSE.

In the following example the model includes the effects of 'Walk', 'Diet', and the

132

interaction between them upon 'Weight'. A special test will be performed in the command TEST about the null hypothesis that 'Walk' has no effect when the interaction between 'Walk' and 'Diet' is considered as error. The MEANS subcommand calculates the mean for each subgroup generated by the 'Walk' and 'Diet' cross classification.

```
MTB > ANOVA 'Weight' = 'Walk'|'Diet';
SUBC> TEST  'Walk'/'Walk'*'Diet';
SUBC> MEANS 'Walk'*'Diet'.


F-test with denominator: Walk*Diet
Denominator MS = 177.72 with   2 degrees of freedom

Numerator    DF       MS       F       P
Walk          1     76.06    0.43    0.580

  Walk Diet    N     Weight
    1    1     3     24.000
    1    2     3    -36.000
    1    3     3    -43.667
    2    1     3     16.667
    2    2     3    -21.667
    2    3     3    -38.333
```

Only the results of the TEST and MEANS subcommands are displayed. The F value of 0.43 and its associated probability of 0.590 does not allow us to reject the null hypothesis that 'Walk' has no effect, when interaction is considered.

7.4 UNBALANCED DATA

<u>Problem</u>

When an experiment is designed, typically the number of cases is the same in each group or cell formed by the combination of categories of the factors or independent variables. TWOWAY is used for such balanced data where interaction is viewed as orthogonal to (independent of) the main effects. However, in non-laboratory research the numbers of observations in the cells are not likely to be equal. For unequal and disproportionate cell frequencies, the orthogonality assumption cannot be maintained.

The following is such an example. We often hear that the rich are getting richer, and the poor are getting poorer. Also that the contemporary youth are comparing their poverty to the perceived affluence of retired professionals. Critics attribute part of the blame to the current tax system.

To find out whether such a claim is valid, a student in a continuing education program took a survey of 18 students in his class. He classified these students according to their age and income. He then asked everyone to calculate the amount of additional tax he or she must pay given a salary raise of $1,000 a year. The results are shown in Exhibit 7.4.

Exhibit 7.4

| | Income group | | |
	$25,000-50,000	75,000-120,000	150,000+
Young	365	420	182
(30-40)	360	433	185
		450	
Old	356	483	182
(50+)	362	470	175
	368	442	160
		430	
		481	

Minitab Solution: The REGRESS Command

Minitab offers two solutions for unbalanced data: (1) the GLM command for the analysis of variance with unbalanced data, and (2) the REGRESS command with the option of using dummy variables. In this section the second method is introduced. The REGRESS command performs a regression analysis, which will be explained in detail in Chapters 9 and 10.

In the following Minitab solution, the READ command enters the data of the three variables simultaneously. To create dummy variables, the INDICATOR command is used. From C2 two dummy variables are generated into C11 and C12. Since C2 ('Age') has two values (1 and 2), C11 will be 1 if an observation has 'Age'=1; otherwise C11 is 0. C12 is 1 if 'Age'=2; otherwise it is 0. Similarly C3 with 3 levels results in 3 dummy variables in C13-C15.

Interaction terms are composed by multiplying nonredundant column values. Since 'Age' has two categories, only one(C11) is essential. The value of C12 is determined if C11 is known. As for 'Income' with its 3 categories, C15 is determined if C13 and C14 are known. Therefore only C11, C13, and C14 are used to produce interaction terms (C21, C22).

The BRIEF command, which will be explained in greater detail in Chapter 9, controls the amount of output from the regression analysis. When BRIEF has the value of 3, the maximum amount of information will be displayed.

After the command keyword, REGRESS, specify the dependent variable ('Tax') followed by the number of predictor variables and a list of predictor variables. In this example there are 5 predictor variables: C11, C13, C14, C21, and C22.

```
MTB > READ C1-C3
DATA > 365 1 1
DATA > 360 1 1
DATA > 356 2 1
DATA > 362 2 1
    . . . . . . . . . . . . . .
DATA > end
MTB > INDICATOR C2 into C11-C12          # Age
MTB > INDICATOR C3 into C13-C15          # Income
MTB > #
MTB > # Create indicator variables for the interaction terms
MTB > #   by multiplying appropriate columns
MTB > LET C21 = C11*C13
MTB > LET C22 = C11*C14
MTB > #
MTB > #  Place the indicator variables in the regression
MTB > #    equation
MTB > BRIEF = 3
MTB > REGRESS 'Tax' on 5 predictors C11 C13 C14 C21 C22


Analysis of Variance

SOURCE          DF            SS          MS          F          p
Regression       5        234048       46810     182.31      0.000
Error           12          3081         257
Total           17        237129

SOURCE          DF        SEQ SS
C11              1           748
C13              1          1018
C14              1        231088
C21              1           135
C22              1          1058
```

Although the command BRIEF=3 produces voluminous output, only the pertinent portions are displayed above.

To produce the analysis of variance table, examine the sequential SS table. The "SEQ SS" for C11 is the sum of squares for the effect of 'Age', and it has only 1 degree of freedom.

Add the sum of squares for C13 and C14 from the "SEQ SS" column, which amounts to 232106. This is the sum of squares for the effect of 'Income'. Also adding the degrees of freedom for C13 and C14, results in 2 degrees of freedom.

135

Sum "SEQ SS" for C21 and C22, which is the sum of squares (1193) for the interaction effect. Sum "DF" for C21 and C22, producing the degrees of freedom (2) for the interaction term.

For the error term, the sum of squares (3081) and its degrees of freedom (12) are directly taken from the regression table.

The mean square terms are found by dividing the sum of squares for each effect by their respective degrees of freedom. The final table is as follows:

```
                       Analysis of Variance

    Source              Df          Sum of Squares      Mean Square
    -------------------------------------------------------------------
    Age                 1                  748              748
    Income              2               232106           116053
    Age*Income          2                 1193              596.5
    Error               12                3081              257


MTB > LET K3=748/257              # F ratio for Age
MTB > PRINT K3
  K3          2.91051
MTB > INVCDF 0.95;
SUBC> F 1 12.
      0.9500     4.7472           # Critical value for Age
MTB > LET K4=116053/257          # F ratio for Income
MTB > PRINT K4
  K4         451.568
MTB > INVCDF .95;
SUBC> F 2 12.
      0.9500     3.8853           # Critical value for Income
MTB > LET K5=596.5/257           # F ratio for interaction
MTB > PRINT K5
  K5          2.32101
MTB > INVCDF .95;
SUBC> F 2 12.                     # Critical value,interaction
      0.9500     3.8853
```

As before we need to calculate F ratios and critical values at a given significance level to test the null hypothesis. With an F statistic of 2.91051 which is smaller than the critical F value of 4.7472 at the 0.05 level of significance, we fail to reject the null hypothesis that age has no impact upon the income tax. On the other hand, 'Income' seems to affect 'Tax' with an F statistic of 451.568, which is much larger than the critical value of 3.8853.

The F ratio for the interaction term is 2.32101, which is smaller than the critical F value of 3.8853. We therefore cannot reject the null hypothesis that there is no interaction at the 0.05 level. However, since the interaction effect is not negligible, the above conclusions must be made with caution.

```
NEW MINITAB COMMANDS

   AOVONEWAY, ONEWAY
   TWOWAY, ANOVA, BRIEF
   REGRESS
```

EXERCISES

1. Test the null hypothesis that the number of units being carried are the same, regardless of the person's marital status.

2. Test the subhypothesis that the married and the single students take the same average number of units.

3. Recode the number of work hours per week into 4 categories: 0 hours, 1 to 10 hours, 11 to 20 hours, more than 20 hours. Determine the effects of this variable as well as the marital status on the number of units being carried.

4. Examine the interaction, if any, of these two variables upon the mean number of units carried.

CHAPTER EIGHT
NONPARAMETRIC TESTS

In This Chapter You Will Learn To Perform Nonparametric Tests for:

1. One sample data: Runs test, Median Test, Wilcoxon Test
2. Two sample data: Mann-Whitney Test
3. K sample data: Friedman Test, Kruskal-Wallis Test, MOOD Test
4. Bivariate relations: Chisquare Test, Rank Correlation

Thus far most of our hypothesis testing was carried out based on the assumption that the population is normally distributed. Frequently, however, there is reason to suspect that the population under study is not normally distributed. Worse yet, we may have no inkling as to its form. In such cases, especially when the sample size is small, nonparametric statistics could be used in place of parametric tests. In general, nonparametric tests require very few assumptions about population distributions and parameters. Another advantage is that they can be applied to nominal or ordinal as well as interval data.

8.1 ONE SAMPLE TEST

8.1.1 The Runs Test

Problem

A new branch of a Japanese electronics company has opened in California. Twelve similar executive positions were to be filled singly every three months. To avoid the criticism of ethnocentrism, each time a position opened up, all available candidates, Americans and Japanese, were reviewed at the same time, and one was selected on the basis of merits disregarding race. The resulting racial background of the 12 executives is listed in Exhibit 8.1 in the order of appointment.

Exhibit 8.1

```
J A J J J A A J J A J A
```

Is the selection of the Americans and the Japanese randomly made? The Runs test examines a sequence of events to test if the order of occurrence is random.

Minitab Solution : The RUNS Command

The RUNS test performs a two-tailed Runs test on the data in specified columns.

Syntax:
```
RUNS test [above and below K] for data in C,...,C
```

After the command keyword, RUNS, specify a criterion value (K) to classify the data into two groups. A run is one or more observations in a row that are greater than K, or one or more observations in succession that are less than or equal to K. If K is not designated, the mean of the column is used.

```
MTB > SET C1
DATA> 1 2 1 1 1 2 2 1 1 2 1 2
DATA> END
MTB > # 1: Japanese   2: American
MTB > RUNS C1

   C1

   K =        1.4167

    THE OBSERVED NO. OF RUNS =    8
    THE EXPECTED NO. OF RUNS =    6.8333
     5 OBSERVATIONS ABOVE K     7 BELOW
  * N SMALL--FOLLOWING APPROX. MAY BE INVALID
              THE TEST IS SIGNIFICANT AT   0.4663
              CANNOT REJECT AT ALPHA = 0.05
```

Since K is not specified, the mean, 1.4167, of C1 is used as a dividing point. The values greater than 1.4167 belong to one group, while the rest is assigned to the second group. The data in this example produce 8 runs as follows:

(1) (2) (1 1 1) (2 2) (1 1) (2) (1) (2)

The expected number of runs, 6.8333, is calculated on the condition of having 7 values that are less than or equal to 1.4167 and 5 values that are greater than 1.4167. The probability of getting as extreme a number of runs (8) or more extreme is found, using a normal approximation. With a significance level of 0.05, the null hypothesis of randomness cannot be rejected.

The normal approximation is generally good if at least 10 observations are greater than K and at least 10 are less than or equal to K. Since the above data do not meet this requirement, a Minitab message is displayed in the output, warning that a normal approximation may be invalid.

8.1.2 The Median Test

Problem

A national survey institute claims that the true median age is 16.5 at which teenagers first experience sexual intercourse. Therefore if anyone thinks it is at a younger or older age at which first sexual encounter occurs, the institute grades the respondent as failing a test on sexual knowledge. Dubious of this claim, a researcher in California polled a sample of 15 college students for the age of their first sexual intercourse. The result is shown in Exhibit 8.2. Can she reject the null hypothesis that half of the teenagers have their first sexual experience beyond the age of 16.5?

With a sample size of 15 it is difficult to assume anything about the underlying distribution. Therefore a conservative approach is to opt for a nonparametric test. Evaluate the null hypothesis using a median test to determine whether the median age of the population is significantly different from 16.5, based on this sample data. Also, construct a 95% confidence interval.

Exhibit 8.2

13 14 16 14 13 12 15 13 14 17 13 14 15 15 18

Minitab Solution: The STEST and SINTERVAL Commands

Minitab provides the STEST, which performs a sign test of the null hypothesis that the true median is K, unless the subcommand ALTERNATIVE is used.

Syntax:
```
STEST [of median = K] on data in C,...,C
ALTERNATIVE
```

After the command keyword, STEST, specify a hypothesized median (K). If K is not designated, K=0 is assumed. STEST performs a separate sign test for each column listed.

Unless the subcommand ALTERNATIVE is used or unless ALTERNATIVE is set to zero, Minitab performs a two-tailed test. The ALTERNATIVE subcommand for a one-tailed test works in the same way as described earlier:

```
ALTERNATIVE = -1    Test the alternative hypothesis (median < K)
ALTERNATIVE = +1    Test the alternative hypothesis (median > K)
```

A paired sign test can also be performed by using a two-tailed STEST on the paired differences of the two samples. The difference scores would, of course, have to be calculated first.

The exact p value is obtained for samples of size 50 or less. For samples larger than 50, the p value is computed using a normal approximation with continuity correction.

```
MTB > SET C1
DATA> 13 14 16 14 13 12 15 13 14 17 13 14 15 15 18
DATA> END
MTB > STEST median=16.5 using the data in C1

   SIGN TEST OF MEDIAN = 16.50 VERSUS  N.E.   16.50

                    N   BELOW   EQUAL   ABOVE   P-VALUE      MEDIAN
      C1           15     13       0       2    0.0074       14.00
```

There are 13 observations below the hypothesized value, 16.5, and 2 observations above. The median age for the first sexual intercourse among the teenagers according to this sample is 14.00. With a p value of 0.0074 the Californian researcher can reject the null hypothesis that the true median age is 16.5 years old.

To construct a confidence interval Minitab provides the SINTERVAL command.

Syntax:

```
SINTERVAL [with K% confidence] on data in C,...,C
```

SINTERVAL calculates three sign confidence intervals. The confidence intervals are based on the binomial distribution for the number of values greater than the median (the signs) of the data. Because of the discrete nature of the procedure, it is seldom possible to achieve the exact level of confidence. Minitab therefore displays two intervals with confidence levels above and below the level specified, and the third interval found by a nonlinear interpolation procedure.

If K is not specified, 95% confidence is used.

```
MTB > SINTERVAL on data in C1

SIGN CONFIDENCE INTERVAL FOR MEDIAN

                         ACHIEVED
         N   MEDIAN    CONFIDENCE   CONFIDENCE INTERVAL   POSITION
   C1   15    14.00      0.8815     (  13.00,    15.00)        5
                         0.9500     (  13.00,    15.00)      NLI
                         0.9648     (  13.00,    15.00)        4
```

To understand better the output of the SINTERVAL command, enter the following program:

```
MTB > SET C3
DATA> 1:15
DATA> END
MTB > SORT C1 C2
MTB > PRINT C3 C2
  ROW    C3    C2

    1     1    12
    2     2    13
    3     3    13
    4     4    13  --------------------
    5     5    13  --------            |
    6     6    14         |            |
    7     7    14         |       96.48%
    8     8    14     88.15%            |
    9     9    14         |            |
   10    10    15         |            |
   11    11    15  --------            |
   12    12    15  --------------------
   13    13    16
   14    14    17
   15    15    18
```

In the output from SINTERVAL, Position 5 means that the first confidence interval (88.15%) goes from the 5th smallest value to the 5th largest one. The 5th smallest value is 13, and the 5th largest value is 15. To make this relationship easier to understand, the sampled values are sorted in ascending order, and printed using the SET and SORT commands. The brackets indicate the intervals are added by the authors.

The position 4 indicates that the interval goes from the 4th smallest value (13) to the 4th largest value (15), which indicates the 96.48% confidence interval. Since the exact 95% level is not attainable, Minitab displays the interpolated confidence intervals to be between 13 and 15.

8.1.3 The Wilcoxon Signed-Rank Test

Problem

While the one-sample median test utilizes only the sign of the difference between each observation and the hypothesized median, the Wilcoxon Signed-Rank Test takes into account not only the sign but the magnitude of each observation relative to the hypothesized median. When such information is available, a test statistic which considers all the information might be expected to give better or more precise results.

Using the same data as before (Exhibit 8.2), test the null hypothesis, and construct a 95% confidence interval by the Wilcoxon Signed-Rank Test.

Minitab Solution: The WTEST and WINTERVAL Commands

The WTEST command performs a one-sample (or a paired sample) Wilcoxon Signed-Rank Test of the null hypothesis that the central value is K.

Syntax:
```
WTEST [of center = K] on data in C,...,C
      ALTERNATIVE
```


If K is not given, K=0 is used. WTEST assumes a random sample from a symmetric distribution, hence the center is the same as the median and the mean. For a paired-sample test, use WTEST on the difference scores with center = 0.

Unless the subcommand ALTERNATIVE is used, Minitab tests the alternative hypothesis that the center is not equal to K.


```
MTB > WTEST center=16.5 data in C1

 TEST OF MEDIAN = 16.50 VERSUS MEDIAN N.E.  16.50

                 N FOR    WILCOXON              ESTIMATED
           N     TEST     STATISTIC  P-VALUE     MEDIAN
  C1       15     15          6.0     0.002       14.00
```


The ESTIMATED MEDIAN (14.00) is the median of all the pairwise (Walsh) averages, $(Y_i + Y_j)/2$, for i less than or equal to j.

N FOR TEST (15) refers to the number of observations which are not equal to K, since WTEST first eliminates cases equal to K. The WILCOXON STATISTIC (6.0) corresponds to the number of Walsh averages (not observations) exceeding the hypothesized center of 16.5, plus one half the number of Walsh averages equal to the hypothesized center, if any.

The P-VALUE is calculated using a normal approximation with the continuity correction. Note that the p value from this test is 0.002, as compared to 0.0074 using the median test. The Wilcoxon Test, which uses the information on the magnitudes as well as the signs of the differences between the observed values and the hypothesized median, results in a smaller Type 1 error in rejecting the null hypothesis when it is true.

A confidence interval for the center is calculated by the WINTERVAL command, using the Wilcoxon Signed-Rank Test procedure.

```
Syntax:   WINTERVAL [with K % confidence] for data in C,...,C
```

```
MTB > WINTERVAL C1

                  ESTIMATED   ACHIEVED
            N      MEDIAN    CONFIDENCE   CONFIDENCE INTERVAL
   C1       15     14.00        95.0    (   13.50,    15.50)
```

Since WINTERVAL assumes a symmetric population, the center equals both the median and the mean. The ESTIMATED MEDIAN(14.00) is the median of all the pairwise averages $(Y_i + Y_j)/2$, where i is less than or equal to j.

The 95% confidence interval [13.5, 15.5] is the range of values, for which the Wilcoxon two-tailed test of Ho:(center = K) is not rejected at the 0.05 level. Because of the discrete nature of the Wilcoxon test statistic, the specified confidence is seldom attained. The closest value therefore is obtained, using a normal approximation with the continuity correction.

8.2 TWO SAMPLE TEST

8.2.1 The Mann-Whitney Test

Problem

The hazards of exercising in polluted air have been well publicized recently. According to a new study, people with coronary artery disease who exercise in air polluted by carbon monoxide run a higher risk of fatal, abnormal heart behavior.

To test partially the validity of such research findings, a physical fitness instructor compared the percentages of carbon monoxide in the blood for joggers on Santa Monica beach and those on downtown L.A. streets. Presumably, a higher level of carbon monoxide in the blood will cause heart problems. The 14 joggers examined are all healthy males in their 30s, who had less than 1 percent carbon monoxide in their blood before the experiment. The carbon monoxide levels after one hour of jogging in the two areas are shown in Exhibit 8.3.

Exhibit 8.3

| Santa Monica | 0.1 | 0.15 | 0.2 | 0.23 | 0.19 | 0.36 | 0.1 |
| L.A. downtown | 0.25 | 0.62 | 0.35 | 0.25 | 0.45 | 0.52 | 0.3 |

Can the fitness instructor reject the null hypothesis that there is no difference in the carbon monoxide levels in the blood after exercising in the cleaner beach air versus in the polluted downtown air of Los Angeles?

The Mann-Whitney Test, or the Rank Sum Test, can be used to compare two independent samples drawn from populations which may not be normally distributed. It computes the sum of ranks for each sample, then tests whether the difference between these sums is significantly different from what is expected under the null hypothesis.

Minitab Solution: The MANN-WHITNEY Command

The MANN-WHITNEY command performs a two sample rank test for the difference between the two population medians. It also calculates point and interval estimates.

Syntax:
```
MANN-WHITNEY [% confidence K] for data in C and C
   ALTERNATIVE
```

If no subcommand is used, a two-tailed test is performed. For instructions on using the subcommand, ALTERNATIVE, see the section on STEST.

The two samples are ranked together, with the smallest observation given rank 1. Then the ranks of the two samples are summed separately. The Mann-Whitney test statistic is calculated in terms of the rank sum of the first or the second sample.

The data from the two samples must be entered in two columns, and specified after the command keyword, MANN-WHITNEY. Without the desired confidence level given, the default 95% is used.

```
MTB > READ C1 C2
DATA>   0.10    0.25
DATA>   . . .
DATA> END
MTB > MANN-WHITNEY C1 C2
```

```
Mann-Whitney Confidence Interval and Test

C1              N =    7      Median =         0.1900
C2              N =    7      Median =         0.3500

Point estimate for ETA1-ETA2 is         -0.1600

95.9 pct c.i. for ETA1-ETA2 is (-0.3700,-0.0600)
W = 32.0

Test of ETA1 = ETA2   vs.   ETA1 n.e. ETA2 is significant
  at 0.0106

The test is significant at 0.0104 (adjusted for ties)
```

The output shows the number of observations and the median for each sample. The median value for the Santa Monica beach joggers is 0.19% carbon monoxide in their blood, while the median for the L.A. downtown joggers is 0.35%.

A point estimate of the difference of the population medians yields -0.16. An approximate 95% confidence interval for the difference in the medians extends from -0.37 to -0.06.

The value W=32.0 is the sum of the ranks for the values of the column number given first in the program command, which in this case is C1.

The initial significance level, 0.0106, of the test is adjusted for ties, resulting in 0.0104. With this value the physical fitness instructor rejects the null hypothesis, and concludes that polluted air makes a significant difference in the carbon monoxide level in the blood.

8.3 K-SAMPLE TEST

8.3.1 The Friedman Test

Problem

Five candidates for state assembly are making their election campaigns. They are asked at news conferences what percentage of the current budget they intend to increase or decrease in five areas: health care, education, housing, environmental control, and crime prevention. The results are shown in Exhibit 8.4 Do these data provide sufficient evidence to conclude that a difference exists in perceptions of the overall priority in budget allocation among the five areas?

Exhibit 8.4

Candidate	Health	Education	Housing	Environment	Crime
A	41	-10	35	15	25
B	20	20	15	15	20
C	15	30	15	30	15
D	35	25	- 5	10	43
E	10	35	42	35	25

For such data, the Friedman Test provides a nonparametric counterpart to the randomized block design analysis of variance. It is an extension of the paired-difference design to compare K population (treatment) means. It tests the null hypothesis that the treatment (the areas for budget allocation) has no effect, that is, the distribution of observations within a block (the candidate) is the same. Unlike the F test, no assumption about the distribution is necessary to use the Friedman test to compare the K populations.

Minitab Solution: The FRIEDMAN Command

The FRIEDMAN command performs a Friedman Test, a nonparametric analysis of a randomized block design.

Syntax:
```
FRIEDMAN data in C, treat. in C, block in C
     [res. into C [fits into C]]
```

After the command keyword, FRIEDMAN, specify the sample data in the first column, the treatment identification values in the second column, and the block numbers in the third column. Additional columns may be listed to store the residuals and the fits (observed values minus the residuals).

In the following example C1 contains the percentage increase or decrease of the budget in various areas (treatment) advocated by each state assembly candidate (block). The treatment numbers are generated into C2 by using the multipliers in the SET command. The block numbers are also produced by the multipliers in the SET command and entered into C3. C4 is specified to store the residuals, and C5 for the fits.

```
MTB > SET C1
DATA> 41 20 15 35 10 -10 20 30 25 35 35 15 15 -5 42
DATA> 15 15 30 10 35 25 20 15 43 25
DATA> END
MTB > SET C2
DATA> (1:5)5
DATA> END
MTB > SET C3
DATA> 5(1:5)
DATA> END
MTB > FRIEDMAN C1 C2 C3 C4 C5
```

```
Friedman test of C1 by C2 blocked by C3

S = 0.60   d.f. = 4   p = 0.963
S = 0.67   d.f. = 4   p = 0.954 (adjusted for ties)

                        Est.      Sum of
          C2      N    Median      RANKS
          1       5    26.000      16.0
          2       5    25.000      16.0
          3       5    22.000      13.5
          4       5    22.000      13.5
          5       5    25.000      16.0

   Grand median   =    24.000
```

The Friedman test ranks the data within each row (block), and provides an analysis of the column (treatment) rank sums. For each treatment, the output shows the number of observations, the estimated median, and the sum of ranks. The overall median is 24.00% increase in budget allocation.

The Friedman's test statistic, S, is approximated by the chisquare distribution with K-1 degrees of freedom where K is the number of treatments. With a p value of 0.954 adjusted for ties, we cannot reject the null hypothesis. Candidates do not seem to discriminate among areas for budget allocation.

8.3.2 The Kruskal-Wallis Test

Problem

Administrative and executive positions had traditionally been monopolized by white males. Although these positions have been relatively more open for women and racial minorities in recent years, conflicts have been reported in the power relations between the female or racial minority bosses and their subordinates. A Personnel Department of a large corporation chose 6 white males, 4 white females, 4 Asian males, and 3 Asian females who are considered equally efficient department heads by the company. Then the Department examined how their subordinates evaluate the effectiveness with which their bosses perform the supervisorial tasks. The scores are shown in Exhibit 8.5. Can the Personnel Department reject the null hypothesis that there is no difference in evaluation among these four groups?

For this type of data an appropriate analytic tool is the Kruskal-Wallis test, which is the nonparametric counterpart to the oneway analysis of variance. Unlike the analysis of variance, the assumption of population normality is unnecessary. This test assumes only that K independent random samples are drawn from continuous distributions with the same shape. The null hypothesis of no difference among the K populations is tested against the alternative that there is a difference.

Exhibit 8.5

White Male	White Female	Asian Male	Asian Female
95	88	75	66
80	90	90	95
75	60	80	80
85	92	86	
79			
82			

Minitab Solution: The KRUSKAL-WALLIS Command

The KRUSKAL-WALLIS command performs a Kruskal-Wallis test.

Syntax:

```
KRUSKAL-WALLIS test for data in C, indices in C
```

After the command keyword, KRUSKAL-WALLIS, specify the column where all the data are stored for all the samples. The second column listed is the column which contains the identification of the K groups (or levels).

```
MTB > READ C1-C2
DATA> 95   1
DATA>  .  .  .
DATA> 80   4
DATA> END
MTB > KRUSKAL-WALLIS C1 C2
```

LEVEL	NOBS	MEDIAN	AVE. RANK	Z VALUE
1	6	81.00	8.5	-0.30
2	4	89.00	10.4	0.62
3	4	83.00	8.7	-0.11
4	3	80.00	8.5	-0.19
OVERALL	17		9.0	

$H = 0.39$ d.f. $= 3$ $p = 0.941$
$H = 0.40$ d.f. $= 3$ $p = 0.941$ (adj. for ties)

* NOTE * One or more small samples

Observations from all the samples are combined, then ranked, with the tied observations being given the average rank. For each level (sample), the number of observations (NOBS), the sample median, the average or mean ranks, and the Z value of the mean rank are printed. The Z values indicate the position of the sample mean rank in relation to the "OVERALL AVE. RANK" of 9.0 in a distribution approximately normal with mean 0 and variance 1.

Under the null hypothesis, the Kruskal-Wallis test statistic (H) can be approximated by a chisquare distribution with K-1 degrees of freedom where K is the number of samples (levels). With a p value of 0.941 the Personnel Department cannot reject the null hypothesis. It appears that the average managerial performance perceived by the lower ranking personnel does not differ significantly whether the boss is white or Asian, male or female.

8.3.3 The MOOD Test

For comparing K samples there is another technique other than Kruskal-Wallis, namely, the MOOD test, which is an extension of the median test for K samples. MOOD tests the null hypothesis that all groups have the same median. The test assumes that independent random samples have been drawn from distributions with the same shape.

The MOOD test uses information about the magnitude of each of the n observations relative to a single number, the median of the pooled samples. As such, it uses less information available than the Kruskal-Wallis test, which takes into account the relative magnitude of each observation when compared to every other observation. Depending on the data and the population distribution, the MOOD test can be more robust than the Kruskal-Wallis test.

Minitab Solution: The MOOD Command

The MOOD test is analogous to the parametric oneway analysis of variance.

```
Syntax:    MOOD data in C, indices in C
               [res. into C [fits into C]]
```

After the command keyword, MOOD, first specify the column which contains sample data for all populations, then the column which provides group identification. The optional third column stores the residuals, and another optional column saves the fits (cell medians).

```
MTB > # Use the same data
MTB > MOOD C1 C2 C3 C4
```

```
Mood median test of C1

    Chisquare = 1.95    df = 3    p = 0.583

                                     Individual 95.0% CI's
C2   N<=   N>   Median   Q3-Q1  ----------+---------+---------+------
 1    4    2     81.0      9.5                     (----+--------)
 2    1    3     89.0     24.5   (-----------------------------+--)
 3    2    2     83.0     12.8                  (-------+------)
 4    2    1     80.0     29.0        (-------------+--------------)
                                ----------+---------+---------+------
                                         70        80        90
Overall median = 82.0
* NOTE * Levels with < 6 obs. have confidence < 95.0%
```

The Mood statistic has a chisquare distribution with K-1 degrees of freedom where K is the number of samples. With a p value of 0.583 the null hypothesis of no difference cannot be rejected. Note that this value, 0.583, is smaller than the p value, 0.941, obtained under the Kruskal-Wallis test.

For each level, the median, the number of observations below and above the overall median, the interquartile range, and the 95% confidence interval are displayed. For example, among the white male department heads (level=1) the median performance score is 81.0. There are 4 observations equal to or below the overall median(82.0), and 2 observations above 82.0. A Minitab warning is displayed that the calculated confidence interval is less than 95% for those levels which have fewer than 6 observations.

8.4 BIVARIATE RELATIONS

8.4.1 The Chisquare Test

Problem

Air bags as an auto safety measure have been growing in popularity. They are normally stored at the steering wheel and inflate rapidly in case of accident. An auto safety agency analyzed accident data to determine the relationship between fatal injury, and the use of an air bag or a seat belt. The results are shown in Exhibit 8.6.

Exhibit 8.6

	Neither	Seat Belt	Air Bag	Both
Fatal injury	38	35	12	14
Nonfatal injury	67	112	43	112

The data displayed above can be viewed as a cross-classification of two nominal variables, which is called a contingency table. The name derives from the fact that if the proportions of the fatally injured vary among various types of safety measures, then the classification of accidents by the type of injury is contingent upon the classification by the safety measure.

The chisquare test is a nonparametric technique to determine whether or not two nominal variables are independent. Using this test, can the auto safety agency reject the null hypothesis that wearing a seat belt and/or being equipped with an air bag makes no difference upon the seriousness of an accident?

Minitab Solution: The CHISQUARE and TABLE Commands

Minitab provides two commands to perform the chisquare test. The first is the CHISQUARE command which is used for data already collapsed into a contingency table. The second is the subcommand CHISQUARE under the TABLE command, which handles raw data. CHISQUARE as a command is dealt with first.

Syntax:
```
CHISQUARE analysis on frequency table in C,...,C
```

The CHISQUARE command performs a chisquare test for independence or association (non-independence) in a contingency table which has already been stored in the worksheet. Up to seven columns can be specified with CHISQUARE.

Data are stored column-wise and entered row-wise according to this command. Thus, the contingency table in Exhibit 8.6 is entered into C1 through C4 for two rows. C1 contains the numbers of fatal and nonfatal accidents of drivers without seat belts or air bags. C2 refers to those with seat belts, C3 with air bags, and C4 with both. The categories of the second classificatory variable, the type of injury, are represented by the first and the second rows of the data entry.

The CHISQUARE command specifies the columns which refer to the categories of one of the categorical variables.

```
MTB > READ C1-C4
DATA> 38   35   12   14
DATA> 67 112   43 112
DATA> END
MTB > CHISQUARE C1-C4
```

```
Expected counts are printed below observed counts

                 C1          C2          C3          C4       Total
        1         38          35          12          14          99
               24.01       33.61       12.58       28.81

        2         67         112          43         112         334
               80.99      113.39       42.42       97.19

Total            105         147          55         126         433

ChiSq =    8.156 +    0.058 +    0.026 +    7.612 +
           2.418 +    0.017 +    0.008 +    2.256 = 20.551
df = 3

MTB > INVCDF 0.95;
SUBC> CHISQUARE df 3.
      0.9500      7.8147
MTB > CDF 20.551 store in K1;
SUBC> CHISQUARE df 3.
MTB > LET K1=1-K1
MTB > PRINT K1
  K1        0.000130355
```

The output includes the cross tabulation with observed and expected frequencies. The chisquare statistic is computed to be 20.551 with 3 degrees of freedom.

Using the INVCDF function, the critical chisquare value at the 0.05 significance level is 7.8147. Since the chisquare statistic based on the data, 20.551, is larger than the critical value, 7.8147, the null hypothesis is rejected.

To obtain an exact p value for the chisquare value of 20.551, CDF is used with 3 degrees of freedom. The resulting probability of obtaining a value as large as or larger than 20.551 is extremely small, 0.000130355; hence the null hypothesis is rejected. The auto safety agency can safely conclude that air bags and/or seat belts do make a difference in the severity of injury in case of accidents.

Suppose the data are available in raw form without being collapsed into a contingency table. Then we must use the TABLE command with the CHISQUARE subcommand. Let us convert the tabulated data into raw data using multipliers in the SET commands. Print the results to confirm the data entry.

```
MTB > SET C1
DATA> 38(1)  67(2)  35(1)  112(2)  12(1)  43(2)  14(1)  112(2)
DATA> END
MTB > SET C2
DATA> 105(1)  147(2)  55(3)  126(4)
DATA> END
MTB > PRINT C1-C2
```

```
ROW   C1   C2

  1    1    1
  2    1    1
  3    1    1
 .  .  .
 38    1    1
 39    2    1
 .  .  .
433    2    4
```

The TABLE command with the CHISQUARE subcommand performs a chisquare test on the raw categorical data.

Syntax:

```
TABLE the data classified by C,...,C

    MEANS               DATA                TOTPERCENTS
    MEDIANS             N                   CHISQUARE
    SUMS                NMISS               MISSING
    MINIMUMS            PROPORTION          NOALL
    MAXIMUMS            COUNTS              ALL
    STDEV               ROWPERCENTS         FREQUENCIES
    STATS               COLPERCENTS         LAYOUT
```

TABLE as explained in Chapter 2 prints oneway, twoway and multiway tables. Due to space limitation, subcommands are not explained here except for CHISQUARE. Use the Help facility to obtain detailed information.

The CHISQUARE subcommand is followed by an argument, output code = K. The value of K determines how many statistics are placed in each cell. The default value is K=1.

K=1   Observed frequency count
K=2   Observed and expected frequency count
K=3   Count, expected count, and standardized residual

With the output code set to 3, the cell contents include the observed frequency count, the expected count, and the standardized residual, where:

Standardized residual = (observed count-expected count)/sqrt(expected count)

The chisquare statistic, 20.551, obtained here by the TABLE command is the same as the one calculated by the CHISQUARE command earlier.

```
MTB > TABLE data in C1 and C2;
SUBC> CHISQUARE output code=3.


    ROWS:  C1        COLUMNS:  C2

                1           2           3           4         ALL

     1         38          35          12          14          99
            24.01       33.61       12.58       28.81       99.00
             2.86        0.24       -0.16       -2.76          --

     2         67         112          43         112         334
            80.99      113.39       42.42       97.19      334.00
            -1.55       -0.13        0.09        1.50          --

   ALL        105         147          55         126         433
           105.00      147.00       55.00      126.00      433.00
              --          --          --          --          --

   CHI-SQUARE =    20.551    WITH D.F. =      3
      CELL CONTENTS  --
                       COUNT
                       EXP FREQ
                       STD RES
```

8.4.2 The Rank Order Correlation

Problem

Do Japanese doctors enjoy as high an occupational prestige as their American counterparts? Do the Japanese view priests differently than the Americans? To answer such questions an occupational counsellor compared the prestige rankings of comparable occupations reported in the United States and Japan. The American scores range from 1 to 100, while the Japanese ratings are between 1 and 30. The results are shown in Exhibit 8.7 Is there a relationship between the two?

In order to measure the degree of association between the two ranked variables, a rank correlation is appropriate. Spearman's rho (rank correlation) is used here because its computation is identical to that for Pearson's correlation coefficient (See Chapter 9) except that it is applied to the ranks as opposed to the actual scores of the variables.

Exhibit 8.7

	U.S.	Japan
State governor	93	26.22
College professor	89	25.44
County judge	87	25.31
Officer of large company	86	24.49
Physician	93	23.03
State government head	87	22.81
Architect	86	20.49
Officer of labor union	75	19.23
Priest	86	17.54
Public school teacher	78	18.27
Policeman	67	12.32
Carpenter	65	9.78
Barber	59	9.54
Truck driver	54	9.07
Road worker	47	5.20
Shoe shiner	33	3.14

Minitab Solution: The RANK and CORRELATE Commands

To calculate Spearman's rho between two columns, C1 and C2, rank both columns of data separately, then execute the CORRELATE command on the columns of ranks. CORRELATE calculates the Pearson correlation coefficients between each pair of columns.

Syntax:
```
CORRELATE the data in C,...,C  [ store in M ]
```

After the command keyword, CORRELATE, specify columns of data. Between each pair of columns correlation coefficients are calculated. For details see Chapter 9.

```
MTB > READ   C1-C2
DATA>    93    26.22
DATA>    89    25.44
DATA>  . . .
DATA> END
MTB > RANK C1 into C3
MTB > RANK C2 into C4
MTB > CORRELATE C3 C4

Correlation of C3 and C4 = 0.941
```

The output shows a very high correlation coefficient, 0.941. The American and Japanese rankings of occupational prestige are monotonically related.

```
NEW MINITAB COMMANDS

    RUNS, STEST, SINTERVAL, WTEST, WINTERVAL
    MANN-WHITNEY
    FRIEDMAN, KRUSKAL-WALLIS, MOOD
    CHISQUARE
    RANK
    CORRELATE
```

EXERCISES

1. Test the hypothesis that the median hours of work is 15 hours per week, using the Median Test and the Wilcoxon Test.

2. Test the hypothesis that GPA is the same between students who work more than 15 hours and those who do not, using the Mann-Whitney Test.

3. Test the hypothesis that GPA is affected by the number of study hours and marital status, using the Kruskal-Wallis Test and the MOOD Test. Recode the number of study hours into three categories.

4. Test the hypothesis that the number of working hours is independent of marital status, using the chisquare test.

5. Compute the rank correlation coefficient between the college-life satisfaction levels of this year and last year.

CHAPTER NINE
CORRELATION AND REGRESSION

```
In This Chapter You Will Learn To:

    1.   Compute correlation coefficients
    2.   Compute regression coefficients
    3.   Construct confidence and prediction intervals
```

This chapter deals with relationships between interval-scale variables. When interest lies in the degree of association between random variables, correlation analysis is used to obtain a summary measure for the relationship. Regression analysis, on the other hand, yields a model for predicting values of the dependent variable from the values of the independent variable.

9.1 CORRELATION COEFFICIENT

Problem

Exhibit 9.1

| | Cost of Living Index (1989) | | |
	Food	Housing	Health
Albany,NY	101.0	116.3	103.0
Beaumont,TX	102.3	75.9	108.5
Chicago,IL	102.4	177.7	109.7
Cleveland,OH	96.4	121.0	106.4
Davenport,IA	99.8	88.3	86.5
Harrisburg,PA	103.4	91.7	100.2
Houston,TX	107.9	80.1	102.7
Lancaster,PA	104.1	89.1	86.5
Milwaukee,WI	99.2	120.5	103.3
Minneapolis,MN	95.3	112.3	102.1
Omaha,NE	87.8	88.2	87.2
Sacramento,CA	103.6	104.8	123.8
Salt Lake City,UT	94.8	84.7	105.6
St Louis,MO	99.7	96.8	99.9
Youngstown,OH	92.5	82.6	84.1

The cost of living varies depending on the region. In some cities things tend to be more expensive, whether it is food, housing, or health care. On the other hand, some cities may have a depressed real estate market, while the grocery bills are very high. To determine correlations among expenditures for food, housing, and health, a random sample of metropolitan areas reported in the Statistical Abstract, 1990 is selected, and their cost of living indices are compared.

The cost of living index measures relative price levels for consumer goods in selected areas for an average standard of living. The national average equals 100, and each city's index indicates a percentage of the national average. For example, a score of 116.3 in Albany indicates that the housing expense is 16.3% above the national average.

Minitab Solution: The CORRELATE Command

The CORRELATE command calculates the (Pearson's) correlation coefficients between each pair of columns.

Syntax:

```
CORRELATE the data in C,...,C [store in M]
```

If more than two columns are specified, every pair of columns is correlated, and a lower half of a correlation matrix is displayed.

If some data are missing, the pairwise deletion method is adopted. That is, the correlations between each pair of columns are calculated using only those rows which do not have one or more missing values.

```
MTB > READ C1-C3
DATA> 101.0    116.3    103.000
DATA> . . .
DATA> END
MTB > NAME C1 'Food' C2 'House' C3 'Health'
MTB > CORRELATE C1 C2 C3

            Food     House
    House   0.086
    Health  0.408    0.391
```

The correlation coefficient between housing and food expenses (0.086) is almost zero. The correlations between food and health (0.408), and housing and health (0.391) costs are not very high either.

Since the correlation coefficient measures linear relationships, a zero coefficient does not necessarily mean two items are not related in any way. They may be related curvilinearly. To examine the relationships visually, PLOT is used as follows:

159

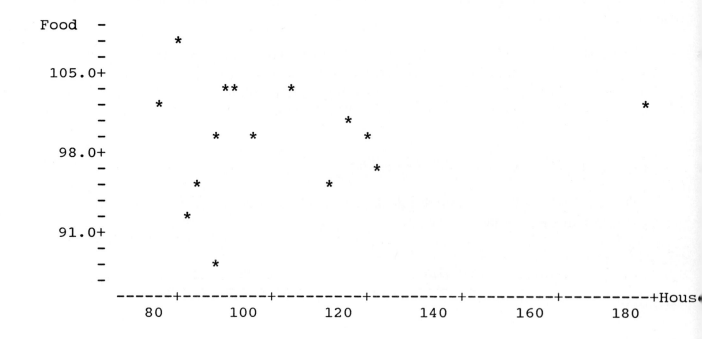
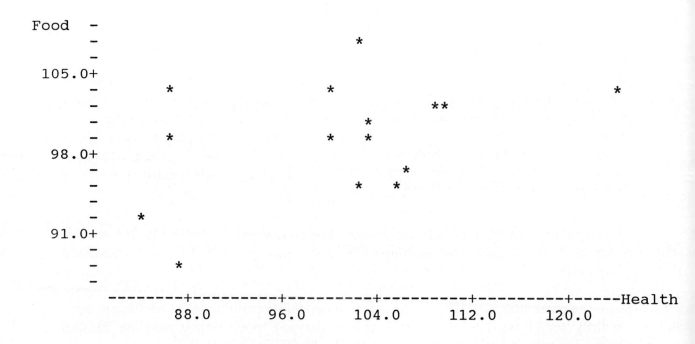

MTB > PLOT C2 C3

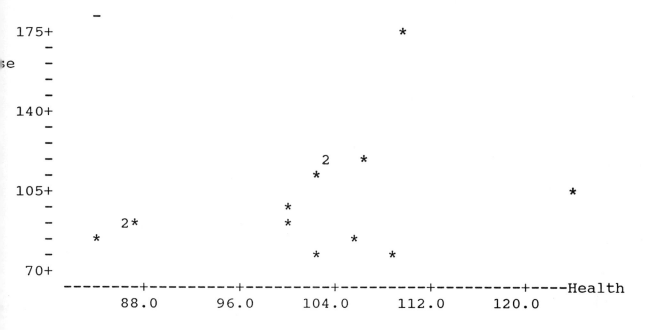

The scatterplots indicate that no special patterns are discernible among the three areas of expenses.

9.2 REGRESSION COEFFICIENT

Problem

In 1990, the year of the budget negotiation, it would be interesting to find out people's attitude toward their elected officials. When Congress appears to be unduly influenced by special interest lobbies, unable to come to any agreement while talking about taxing the rich but also about levying duties on gasoline, beer and cigarettes which would disproportionately affect the poor, what would Americans say to re-electing the incumbents in the November election? Do people's attitude vary depending on the economic status of the state in which they reside?

Suppose we abstract figures for the economic status of the 50 states and the District of Columbia from the Statistical Abstract, 1990 as shown in Exhibit 9.2. The economic affluence versus poverty of each state is measured by two variables: personal income per capita ('Income') and the percentage of residents who are public aid recipients ('Welfare').

Then each state is polled as to people's willingness to dump the incumbents who are up for re-election. This third variable is measured by the number of voting residents per 2,000 voting residents who indicate that they will vote one or more incumbents out of the office. We will call this variable 'No_Incum'. You realize, of course, that this column of data in Exhibit 9.2 is fictitious, since no such poll is conducted.

Exhibit 9.2

ROW	State	Income	No-Incum	Welfare
1	Maine	15106	157	6.1
2	New Hampshire	19434	148	1.7
3	Vermont	15302	142	5.2
4	Massachusetts	20816	620	5.9
5	Rhode Island	16892	397	5.8
6	Connecticut	23059	455	4.2
7	New York	19305	1097	7.7
8	New Jersey	21994	583	5.2
9	Pennsylvania	16233	362	5.9
10	Ohio	15536	452	7.2
11	Indiana	14924	380	3.7
12	Illinois	17575	810	7.0
13	Michigan	16552	742	8.4
14	Wisconsin	15524	214	7.1
15	Minnesota	16674	290	4.6
16	Iowa	14662	257	4.7
17	Missouri	15452	553	5.5
18	North Dakota	12833	59	3.3
19	South Dakota	12755	114	3.9
20	Nebraska	14774	273	3.5
21	Kansas	15759	365	3.7
22	Delaware	17661	452	4.1
23	Maryland	19487	807	4.9
24	Dist.of Columbia	21389	1922	10.6
25	Virginia	17675	299	3.8
26	West Virginia	11735	131	8.1
27	North Carolina	14304	502	5.1
28	South Carolina	12926	741	5.8
29	Georgia	15260	665	6.4
30	Florida	16603	1118	4.1
31	Kentucky	12822	330	7.0
32	Tennessee	13873	533	6.5
33	Alabama	12851	559	6.3
34	Mississippi	11116	325	11.1
35	Arkansas	12219	423	6.0
36	Louisiana	12292	717	9.1
37	Oklahoma	13323	435	5.0
38	Texas	14586	653	4.7
39	Montana	12886	123	4.6
40	Idaho	12665	235	2.6
41	Wyoming	13609	314	3.3
42	Colorado	16463	473	3.9
43	New Mexico	12488	658	5.7
44	Arizona	14970	610	3.8
45	Utah	12193	243	3.1
46	Nevada	17511	781	2.6
47	Washington	16473	466	5.7
48	Oregon	14885	546	4.1
49	California	18753	930	8.7
50	Alaska	19079	523	4.6
51	Hawaii	16753	257	4.9

Minitab Solution: The REGRESS Command

The REGRESS command performs a regression analysis.

Syntax:

```
REGRESS C on K predictors C,...,C
[put stand. residuals in C [fits in C]]

NOCONSTANT    XPXINV    COOKD     VIF
WEIGHTS       RMATRIX   DFITS     DW
MSE           RESIDS    HI        PURE
COEF          TRESIDS   PREDICT   XLOF
```

REGRESS is followed by the dependent variable column (C), the number of predictors (K), and the column locations of the predictor variables. If an additional column is specified, the standardized residuals are stored. See paragraph on the RESIDS subcommand for an explanation of standardized residuals. If another additional column is indicated, the fitted or predicted values are stored in it. For samples of fitted or predicted values, see the output under the "Fit" column of "Unusual Observations".

To control the amount of printed output, use the command BRIEF described in Chapter 7 and also later in this section.

Only selected subcommands are described here. For further information, use the Help facility.

The NOCONSTANT subcommand fits the equation without the constant.

The PREDICT subcommand computes predicted values of the dependent variable for given values of predictors. The arguments on PREDICT should be all columns containing predictor variables or all constants. The number of arguments on PREDICT should be the same as the number of predictor variables. In addition, PREDICT computes the standard deviations, confidence intervals, and prediction intervals for specified values. See Section 9.3 for the use of this subcommand.

The RESIDS subcommand computes the residuals which are the differences between the observed and predicted values, and stores them in a specified column. The standardized residuals, which are stored as an option at the command line, are the residuals divided by their standard deviations.

The TRESIDS subcommand stores studentized residuals. These are similar to the standardized residuals, except that the calculations are done with the i-th observation omitted from the data set. Therefore the i-th observation cannot influence the calculation, which permits unusual Y values to stand out clearly.

163

The BRIEF Command

To control the amount of output from REGRESS, use the BRIEF command.

Syntax:
> BRIEF [with output code = K] for commands that follow

The larger the K value, the more output will be produced. The default is BRIEF 2.

K = 1   The regression equation, table of coefficients, S, R-squared, R-squared adjusted, the first part of the analysis of variance table

K = 2   Additionally, the second part of the analysis of variance table, the unusual observations in the table of fits, residuals

K = 3   Additionally, the full table of fits and residuals

First, let us examine if the income level helps to predict willingness to vote out incumbents. Following the command keyword, REGRESS, the dependent variable 'No_Incum' is specified. The integer 1 indicates that there is one predictor variable, which is then identified as 'Income'.

```
MTB > READ C1-C3
DATA> 15106    157        6.1
DATA> . . .
DATA> END
MTB > NAME C1 'Income' C2 'No_Incum'
MTB > REGRESS 'No_Incum' 1   'Income'
```

```
The regression equation is
No_Incum = - 298 + 0.0506 Income
```

Predictor	Coef	Stdev	t-ratio	p
Constant	-298.2	231.6	-1.29	0.204
Income	0.05056	0.01454	3.48	0.001

s = 289.9   R-sq = 19.8%   R-sq(adj) = 18.2%

Analysis of Variance

SOURCE	DF	SS	MS	F	p
Regression	1	1016683	1016683	12.10	0.001
Error	49	4118131	84043		
Total	50	5134813			

```
Unusual Observations

Obs.   Income   No_Incum      Fit  Stdev.Fit    Residual     St.Resid
  6     23059      455.0    867.7      114.6      -412.7       -1.55 X
  8     21994      583.0    813.9      100.3      -230.9       -0.85 X
 24     21389     1922.0    783.3       92.3      1138.7        4.14R
 30     16603     1118.0    541.3       42.7       576.7        2.01R
```

R denotes an obs. with a large st. resid.
X denotes an obs. whose X value gives it large influence.

The default (BRIEF 2) output is shown above. The regression equation is given as:
No_Incum = -298 + 0.0506 Income

The regression coefficient is 0.0506. To understand why the regression coefficient appears to be such a small value and is yet significant (see paragraph below), look at Exhibit 9.2 again. Note that the income values are large relative to the No_Incum data. It means that although a relationship may exist between the two, a relatively large change in income could be expected to lead to a relatively small change in No_Incum rate. This is exactly what the coefficient 0.0506 tells us, that as the average income increases by $1, the No_Incum rate goes up by 0.05 per 2,000 residents. Remember that the regression coefficients in the equation are non-standardized, hence they vary depending on the unit of measurement. The least-square equation would not be the same if income is measured in thousands of dollars rather than in dollars, or the No_Incum rate is computed per 10,000 instead of 2,000 resident population. A change of scale will affect the slope of the line, but not the significance of the results.

The table output below the regression equation displays the standard deviation and the t ratio together with the coefficient for each predictor. With a t statistic of 3.48 and its associated probability of 0.001, we can reject the null hypothesis that the regression coefficient is zero. This means that states with higher incomes are more likely to want to dump the incumbents.

The estimated standard deviation about the regression line, denoted by s, equals 289.9. The square of this is the same as the MS Error (84043).

R-squared (the coefficient of determination) is 19.8%, which means that less than 20% of the variation in No_Incum rates is explained by the income level. Since the R-squared is larger for a larger number of predictors in the regression equation, the adjusted R-squared is computed, which takes into account the degrees of freedom. This (18.2%) is an approximately unbiased estimate of the population R-squared. The result certainly confirms the notion that anti-incumbent sentiment runs high in more affluent states. Be very careful, however, about drawing any conclusion concerning individuals based on group (state) data.

The second table in the output is the analysis of variance. The MS mean square due to the regression model is 1016683, while the MS due to error is 84043. The ratio

MS(regression)/MS(error) has an F distribution with 1 and 49 degrees of freedom. The F statistic, 12.10, and its associated probability of 0.001 lead us to reject the null hypothesis that this regression model accounts for nothing. The predictive power of income level may be small (18.2%), but it is not a random occurrence.

Below the analysis of variance table is a list of unusual observations. The fourth column, 'Fit', is the predicted No_Incum rate given an income level using the regression equation. The fifth column, 'Stdev.Fit' is the estimated standard deviation of the fitted value.

'Residual' is the difference between the actual No_Incum rate and the predicted No_Incum rate. 'St.Resid' is the standardized residual for the i-th observation. When the subcommand RESIDS is used, 'St.Resid' is computed for every Y, not just the unusual cases, and stored. Standardized residuals over 2 are considered large, since the variance is 1. Minitab attaches 'R' to an observation which has a standardized residual over 2. According to the output, observations 24 and 30 have standardized residuals over 2. Look within the 'Unusual Observations' section and under the 'St.Resid' column for the R symbol.

The symbol 'X' appears in the same column where the 'R's' are attached. It is given to an observation which has a large influence in estimating the regression coefficient. Generally, an 'X' denotes an observation with 'Income' away from the center. Minitab warns that Observations 6 (Connecticut) and 8 (New Jersey) with high incomes ($23,059 and $21,994) have disproportionate impact on the calculation of the slope.

Let us examine visually whether or not the two variables are linearly related since the regression equation adopts a linear model.

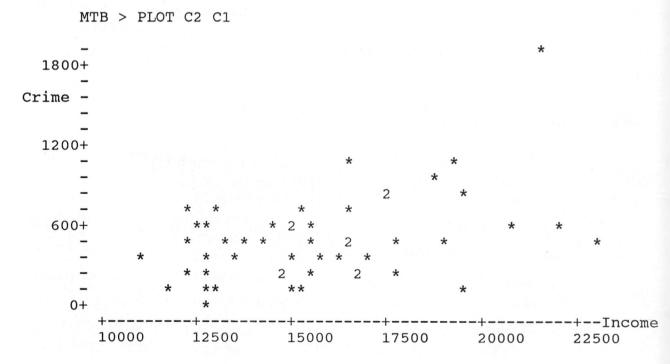

The scatterplot by the PLOT command shows that when the four unusual cases, those to the left of the graph, are eliminated, dots are concentrated in the lower right quadrangle. The line which fits these dots best has a smaller slope than the line which takes into account the extreme cases.

9.3 CONFIDENCE AND PREDICTION INTERVALS

Problem

It has been shown above that the more affluent states are more willing to vote out political incumbents. Would the results remain the same if affluence of the state is measured by the number of welfare recipients in each state? Is the percentage of public aid recipients in the state a good predictor of the attitude toward incumbents of that state?

In addition to testing this hypothesis, construct 95% confidence intervals for the expected values of No_Incum rates in the population for given percentages of the welfare recipients.

In many situations an interval for an actual value of Y, rather than an average or expected value of Y, for a value of X is needed. For example, you may want to estimate the No_Incum rate for a state such as California with 8.7% of welfare recipients. In this case, what is estimated is not a parameter but the value of a random variable. Its probabilistic range is called a prediction interval rather than a confidence interval.

In estimating a single value, the margin of error is larger than in estimating an expected or average value in the population. While a confidence interval deals with the variability in estimating the true mean, a prediction interval contains another source of error, that is, the variability of the individual value about the mean value.

Minitab Solution: The REGRESS Command

The worksheet is sorted by the predictor variable, 'Welfare', so that the confidence and prediction intervals will be displayed in ascending order of the 'Welfare' values. The SET command creates ID numbers for the 50 states and the District of Columbia.

The command BRIEF with value 3 displays the full table of predicted values and residuals, while the default output (BRIEF 2) provides a table only for unusual observations.

At the command line standardized residuals are stored in C17, and predicted values in C18. The subcommand TRESIDS stores the studentized residuals in C15. The subcommand PREDICT prints a table containing the fitted Y values, standard deviations of the fitted Y values, their 95% confidence intervals, and their 95% prediction intervals.

167

```
MTB > SET C4
DATA> 1:51
DATA> END
MTB > SORT C3 C1 C2 C4 C5-C8
MTB > NAME C5 'S_Well' C6 'S_Income' C7 'S_NoIncu' C8 'ID'
MTB > BRIEF 3
MTB > REGRESS 'S_NoIncu' 1 'S_Well' resid C17  predict C18;
SUBC> PREDICT 'S_Well'.
SUBC> TRESIDS C15.
```

The regression equation is
S_NoIncu = 103 + 72.2 S_Well

Predictor	Coef	Stdev	t-ratio	p
Constant	103.4	119.8	0.86	0.392
S_Well	72.21	20.79	3.47	0.001

s = 290.0   R-sq = 19.8%   R-sq(adj) = 18.1%

Analysis of Variance

SOURCE	DF	SS	MS	F	p
Regression	1	1014278	1014278	12.06	0.001
Error	49	4120536	84093		
Total	50	5134814			

Obs.	S_Well	S_NoIncu	Fit	Stdev.Fit	Residual	St.Resid
1	1.7	148.0	226.2	87.4	-78.2	-0.28
2	2.6	235.0	291.2	71.3	-56.2	-0.20
3	2.6	781.0	291.2	71.3	489.8	1.74
. . .						
15	4.1	1118.0	399.5	49.0	718.5	2.51R
. . .						
48	8.7	930.0	731.7	79.3	198.3	0.71
49	9.1	717.0	760.5	86.6	-43.5	-0.16
50	10.6	1922.0	868.8	115.1	1053.2	3.96RX
51	11.1	325.0	905.0	124.9	-580.0	-2.22RX

R denotes an obs. with a large st. resid.
X denotes an obs. whose X value gives it large influence.

Fit	Stdev.Fit	95% C.I.		95% P.I.	
226.2	87.4	(50.5,	401.8)	(-382.6,	835.0)
291.2	71.3	(147.8,	434.6)	(-309.1,	891.4)
291.2	71.3	(147.8,	434.6)	(-309.1,	891.4)
327.3	63.1	(200.5,	454.1)	(-269.2,	923.8)
341.7	60.0	(221.2,	462.2)	(-253.5,	936.9)
. . .					
399.5	49.0	(300.9,	498.0)	(-191.7,	990.6)
406.7	47.9	(310.4,	503.0)	(-184.1,	997.5)
435.6	44.1	(347.0,	524.1)	(-154.0,	1025.2)
435.6	44.1	(347.0,	524.1)	(-154.0,	1025.2)
435.6	44.1	(347.0,	524.1)	(-154.0,	1025.2)
. . .					
515.0	41.0	(432.6,	597.5)	(-73.7,	1103.7)
522.2	41.4	(439.1,	605.4)	(-66.5,	1111.0)
522.2	41.4	(439.1,	605.4)	(-66.5,	1111.0)
529.5	41.8	(445.4,	613.5)	(-59.4,	1118.4)
529.5	41.8	(445.4,	613.5)	(-59.4,	1118.4)
. . .					
688.3	68.9	(549.8,	826.9)	(89.2,	1287.4)
710.0	74.1	(561.1,	858.8)	(108.4,	1311.6)
731.7	79.3	(572.2,	891.1)	(127.3,	1336.0)
760.5	86.6	(586.5,	934.6)	(152.2,	1368.9)
868.8	115.1	(637.6,	1100.1)	(241.8,	1495.9) X
905.0	124.9	(654.0,	1155.9)	(270.3,	1539.6) X

X   denotes a row with X values away from the center

The regression equation using the welfare recipients as a predictor is given as:
S_NoIncu = 103 + 72.2 S_Well

The regression coefficient 72.2 is significant at the 0.001 level. Therefore we reject the null hypothesis that welfare rates have no bearing upon the attitude toward incumbents. As the percentage of welfare recipients increases in the state, so does the willingness to vote out incumbents. This model explains 18.1% of the variation in the No_Incum rates, approximately the same percentage as the model using personal income level as predictor.

The last table shows the confidence and prediction intervals arranged in ascending order of 'Welfare' values. Examine the first observation, which is New Hampshire. The actual No_Incum rate is 148.0 per 2,000 and the welfare recipients constitute 1.7% of the residents. Using the regression model derived from the data on 50 states and DC, the predicted No_Incum rate for New Hampshire, or any other state with 1.7% welfare recipients, is 226.2 per 2,000. The 95% confidence interval for the expected No_Incum rate for this percentage of welfare recipients is between 50.5 and 401.8 per 2,000. The interval width is 351.3.

Note how the interval width varies from Observation 1 through 51. The width of the interval band along the regression line grows larger as it moves away from the center. This means that the regression equation calculated from a sample is best as an estimate of the population regression equation when X=mean of X. The interval is the smallest at that point. The farther one goes in either direction from the center (the mean welfare percentage), the greater the expected inaccuracy.

The predicted No_Incum rate of 226.2 per 2,000 for New Hampshire may be used to predict the No_Incum rate of some other randomly drawn sample that has the same 1.7% of welfare recipients. The 95% prediction interval for this particular state is between -382.6 and 835.0 per 2,000. As expected, the width of the prediction interval is larger than that of the confidence interval due to the greater variation in error in estimating the former.

Let us examine visually how well the prediction model fits the actual data. The following is the scattergram of the predicted and actual No_Incum rates, produced by PLOT. If the fit is good, the dots should be on a diagonal line. With an adjusted R-squared of 18.1%, the scatter plot does not show a very good fit. There may be other factors which influence attitude toward incumbents other than the economic status of the states.

```
MTB > NAME C18 'Predict'
MTB > PLOT 'Predict' vs 'S_NoIncu';
SUBC> XLABEL 'Actual No_Incumb';
SUBC> YLABEL 'Predict No_Incumb'.
```

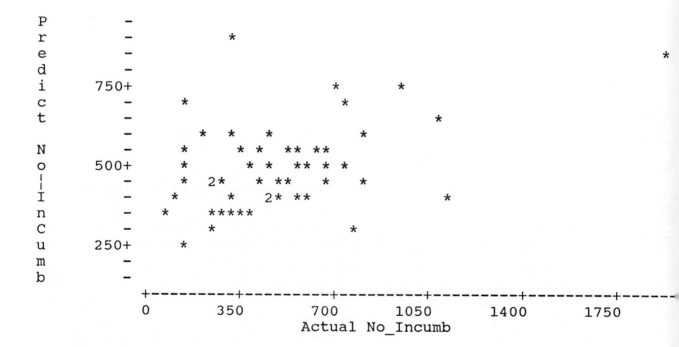

```
New Minitab Commands

    CORRELATE
    REGRESS,  BRIEF
```

EXERCISES

1. Calculate Pearson's r for the number of work hours and the number of units carried.

2. Calculate the regression equation for the same data.

3. Calculate the confidence and prediction intervals for 3 selected values of the number of work hours.

CHAPTER TEN
MULTIPLE REGRESSION

In This Chapter You Will Learn To:

1. Build a regression model
2. Simplify the model
3. Test assumptions for regression analysis
4. Transform nonlinear data for linear regression analysis
5. Perform regression analysis with dummy variables

In Chapter 9 linear regression with one independent variable was discussed. This chapter begins by building a linear regression model with more than one predictor, then techniques for simplifying the model are presented. Additional topics covered are the testing of assumptions for regression analysis, and the handling of nonlinear data and categorical variables.

10.1 BUILDING A REGRESSION MODEL

Problem

There is usually no single cause for an outcome. Regression analysis therefore must include multiple predictor variables. For example, why do some people save money, while others spend every dollar they earn? Do people with higher income also have greater bank accounts? Compared to the rest of the world, Americans are known as spendthrifts.

The following is the data on the saving behavior of a random sample of 23 males between the ages of 20 and 65. Savings are calculated as a percentage of an individual's income. The independent variables are: annual personal income ('Income'), assets ('Asset'), number of children ('Children'), age ('Age'), annual expenditure on leisure-time activities ('Leisure'), annual loan payments ('Loan'), and number of years of schooling ('Educate').

When faced with many factors that seem to affect a dependent variable, the first task is to estimate the coefficients for a regression equation in such a way that the deviation of the actual data from the regression model is minimized. The next task is to verify the validity of the model by testing the null hypothesis that all the regression coefficients are zeroes. If the null hypothesis can be rejected, which validates the model as a whole, then the third step is to pinpoint the predictive power of each independent variable. This is done by examining the slope of each independent variable with the null hypothesis that it equals zero.

Exhibit 10.1

Save(%)	Income	Asset	Children	Leisure	Loan	Educate	Age
41.0	60000	300000	0	3000	0	23	55
10.0	50000	50000	3	5000	15000	18	46
5.0	50000	750000	0	20000	25000	20	43
2.0	50000	100000	4	10000	8000	20	59
22.0	88000	250000	1	10000	5000	18	60
3.0	65000	700000	2	50000	1000	18	62
2.0	42000	55000	3	10000	10000	16	34
1.0	38000	25000	4	7500	7500	16	32
15.0	55000	100000	1	85000	10000	16	48
0.0	34000	10000	1	5000	10000	13	28
1.0	38000	15000	3	10000	10000	13	31
2.5	43000	75000	2	7500	12000	16	45
0.7	28000	5000	5	5000	10000	13	36
1.3	31000	2500	4	3300	7500	11	33
0.0	18000	1000	3	3000	8000	11	26
39.0	58000	250000	0	8000	5000	18	46
12.0	53000	50000	4	5000	15000	20	44
3.0	32000	45000	3	6500	9500	14	35
2.3	29000	2000	3	3000	15000	14	27
0.0	27500	1000	5	5000	9000	12	24
33.0	56000	350000	0	10000	0	18	43
1.0	28500	1000	4	1000	10000	12	28
6.0	39000	9000	2	1000	3000	14	34

Minitab Solutions: The REGRESS Command

```
MTB > READ C1-C8
DATA>  41.0  60000  300000   0  3000  0   23    55
DATA>   . . .
DATA> END
MTB > NAME C1 'Save(%)' C2 'Income' C3 'Asset' C4 'Children'
MTB > NAME C5 'Leisure'
MTB > NAME C6 'Loan' C7 'Educate' C8 'Age'.
MTB > REGRESS C1 7 C2-C8 C18 C19;
SUBC> TRESIDS C11.
```

```
The regression equation is
Save(%)= -2.0 + 0.000227 Income - 0.000014 Asset
            - 3.73 Children - 0.000041 Leisure - 0.000823 Loan
            + 2.20 Educate - 0.374 Age
```

```
Predictor         Coef        Stdev      t-ratio          p
Constant         -2.04        11.02        -0.19       0.856
Income       0.0002270    0.0002366         0.96       0.353
Asset       -0.00001381   0.00001100        -1.25      0.229
Children        -3.728        1.455        -2.56       0.022
Leisure     -0.0000413    0.0001016        -0.41       0.690
Loan        -0.0008230    0.0003281        -2.51       0.024
Educate         2.1977       0.9735         2.26       0.039
Age            -0.3741       0.3765        -0.99       0.336

s = 7.313       R-sq = 77.5%      R-sq(adj) = 66.9%

Analysis of Variance

SOURCE          DF            SS           MS          F          p
Regression       7       2757.33       393.90       7.37      0.001
Error           15        802.23        53.48
Total           22       3559.55

SOURCE          DF        SEQ SS
Income           1       1467.51
Asset            1          6.79
Children         1        702.77
Leisure          1        116.72
Loan             1        187.03
Educate          1        223.71
Age              1         52.80

Unusual Observations
Obs.    Income    Save(%)       Fit    Stdev.Fit    Residual  St.Resi
  9      55000      15.00      10.80       7.08        4.20      2.31R
 16      58000      39.00      25.58       3.14       13.42      2.03R

R denotes an obs. with a large st. resid.
```

At the top of the output is the regression equation:

Save(%) = - 2.0 + 0.000227 Income - 0.000014 Asset
      - 3.73 Children -0.000041 Leisure - 0.000823 Loan
      + 2.20 Educate - 0.374 Age

The linear model with seven independent variables explains 66.9% (adj R-sq) of the variation in the percentage of saving. An F statistic of 7.37 with a significance level of 0.001 results in the rejection of the null hypothesis. The model as a whole is therefore validated.

Below the regression equation is a list of predictors with their coefficients, standard deviations, t ratios, and p values. Using a significance level of 0.05, 'Children', 'Loan', and 'Educate' meet the criterion for rejecting the null hypothesis of no impact upon saving.

Below the analysis of variance table is the 'SEQ SS' for each predictor. The sequential sum of squares represents the reduction in the sum of squared error (SSE) that results from adding the independent variable to the regression equation which includes all predictors specified before it in the REGRESS command. For example, the 'SEQ SS' for 'Children' is the sum of squares for the effect of adding 'Children' to the equation which already includes 'Income' and 'Asset', and it has 1 degree of freedom. The degrees of freedom associated with the variable is one unless the model is not full rank.

A table of unusual observations includes those which are outliers or which have a large influence. Observations 9 and 16 are marked by R's which denote that their standardized residuals are large.

10.2 EXAMINING THE NATURE OF THE DATA

10.2.1 Homoscedasticity

The linear regression model assumes equality of variances. This assumption can be examined by studying a scatterplot of the predicted values and the residuals. If the spread of the residual increases or decreases with the values of the independent variables or with predicted values, then the variance of Y is not the same for every value of X and the assumption is violated.

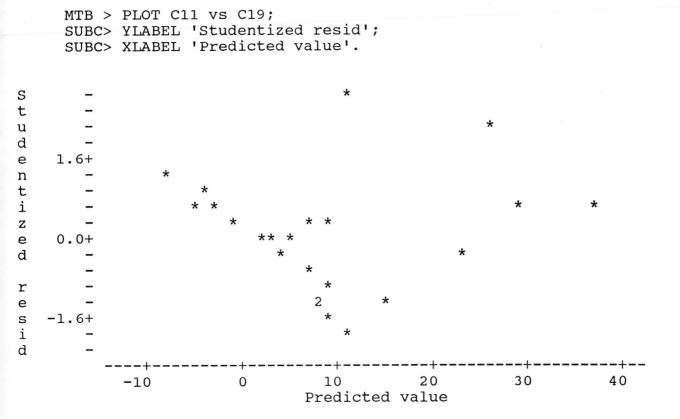

Studentized residuals and predicted values are plotted above. The plot indicates that the residuals for negative predicted values tend to be positive, while the residuals for the positive predicted values are randomly distributed. However, the overall pattern of residuals does not show systematic increase or decrease with particular predicted values; hence the equality of variance assumption is not violated.

10.2.2 Normality Assumption

To test the assumption that residuals are distributed normally, you can compare the observed distribution of residuals to that expected under the assumption of normality. Thus the normality assumption is tested by plotting the NSCORES, namely the normal scores of the standardized residuals against the standardized residuals. The NSCORES are transformations of the standardized residuals into standard normal distribution values or z scores. If the observed and expected distributions are identical, a straight line should result. Since the plot is approximately linear, the normality assumption is tenable.

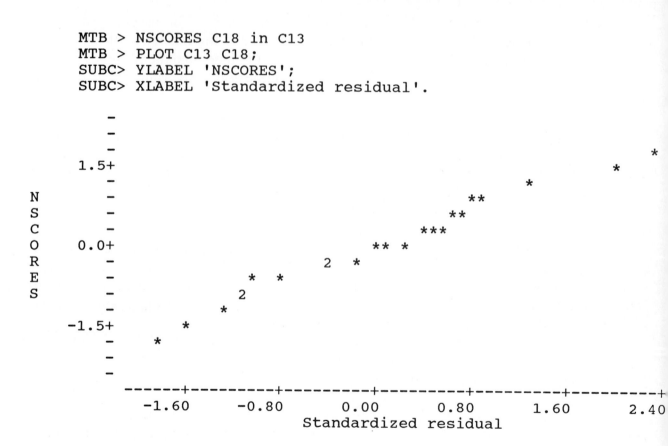

10.2.3 Multicollinearity

In building a parsimonious regression model, the problem of multicollinearity among independent variables must be addressed. The inclusion of independent variables which are highly correlated causes redundancy in the model. To detect such multicollinearity, CORRELATE is executed.

The table indicates a high correlation among 'Age', 'Income', and 'Education'. Older people are more likely to have high education and high income. Thus, you might consider deleting one or more of the highly correlated variables, or combining them into a composite variable as the next step of analysis.

```
MTB > CORRELATE C1-C8

          Save(%)  Income   Asset Children Leisure   Loan  Educate
Income     0.642
Asset      0.345   0.592
Children  -0.695  -0.593  -0.609
Leisure    0.062   0.359   0.379  -0.287
Loan      -0.455  -0.272  -0.015   0.188  -0.029
Educate    0.625   0.767   0.600  -0.513   0.176  -0.047
Age        0.507   0.876   0.602  -0.438   0.408  -0.272   0.802
```

10.3 SIMPLIFYING THE MODEL: STEPWISE REGRESSION

Problem

An original model can be made more parsimonious by eliminating variables which do not show high predictive power. Deletion of redundant, highly-correlated variables is another way toward simplification, as discussed in conjunction with multicollinearity. To simplify a model, REGRESS can be used as above, and any variable whose t probability does not meet a predetermined level can be deleted.

Alternatively, Minitab offers a stepwise regression procedure which is useful for exploratory model building. STEPWISE selects independent variables in three steps: (1) a variable that meets the criteria is entered into a model, (2) the variables already in the model are examined, and (3) any variable that does not meet the criteria is deleted. Let us simplify the saving behavior model by using STEPWISE.

Minitab Solution: The STEPWISE Command

Stepwise regression can be used to identify a subset of predictors from a large collection of variables. The STEPWISE command performs a stepwise regression analysis with three methods: stepwise, forward selection, and backward elimination. See Section 10.7 for more details on this command.

The command keyword, STEPWISE, is followed by the specification of the dependent variable, the number of independent variables, and a list of independent variables.

Without a subcommand, the default stepwise selection is adopted. According to this method, at each step an F statistic is calculated for each predictor in the equation. If any predictor has an F value of less than 4, the default value, then the variable with the smallest F is removed. A new regression equation with one less predictor is calculated and the results are printed. However, if every predictor within the equation has an F value of over 4, then an F statistic is calculated for each predictor not in the equation. The variable with the largest F is then added to the equation, provided that its F value exceeds the F criterion for entry, which also has a default value of 4. The procedure continues until no more predictors can be added.

```
MTB > STEPWISE C1 7 C2-C8

    STEPWISE REGRESSION OF Save(%) ON 7 PREDICTORS, WITH N = 23

        STEP        1         2         3
    CONSTANT   22.3373   27.9852    0.6717

    Children     -5.5      -5.0      -3.4
    T-RATIO     -4.43     -4.34     -2.89

    Loan                -0.00077  -0.00082
    T-RATIO               -2.31     -2.79

    Educate                         1.50
    T-RATIO                         2.65

    S            9.36      8.52      7.47
    R-SQ        48.28     59.19     70.23
```

For each step STEPWISE prints the constant term (unless the NOCONSTANT command is in effect), the coefficient and t ratio for each predictor in the model, standard deviation(s), and R-square.

At step one 'Children' was entered. The regression equation is:

Save(%) = 22.3373 - 5.5 Children

If you have no child, you are expected to save 22.3%, while with one child your saving will drop to 16.8%. The t value for 'Children' is -4.43 with 21 (=23-2) degrees of freedom. Using the 0.05 level of significance, the INVCDF command returns a critical t value of 1.72 for 21 degrees of freedom. Since the absolute value of -4.43 is larger than 1.72, we reject the null hypothesis that 'Children' has no effect on saving. This one predictor model accounts for 48.28% of the variation in saving behavior.

```
MTB > INVCDF 0.95;
SUBC> T 21.
      0.9500     1.7208
```

At step two the equation contains two predictors as follows:

Save(%) = 27.98 - 5.0 Children - 0.00077 Loan

This model explains 59.19% of the variation in the amount of savings. The model says that the more children you have and the more loans you make, the less likely you are to save. If you have 3 children and have $10,000 of loan payments a year, your predicted savings will be 5.28%. Regression coefficients for 'Children' and 'Loan' are both significant at the .05 level with the t values of -4.43 and -2.31 respectively.

At step three 'Educate' is added into the equation as follows:

Save(%) = 0.6717 - 3.4 Children -0.00082 Loan + 1.50 Educate

Education has a positive effect upon saving, while children and loans deter saving. This model explains 70.23% of the variation in the saving pattern. The regression analysis stops at this point because no more predictors can be added to the model based on the criteria discussed above.

10.4 SIMPLIFYING THE MODEL: BREG

There is another Minitab command, BREG, which is useful in simplifying a model. While STEPWISE includes any number of predictors so long as they meet the criteria, BREG selects the best subsets of <u>each size</u>.

<u>Minitab Solution</u>: The BREG Command

The BREG command performs regression analysis on all possible combinations of predictors, then selects the best subsets of each size according to the maximum R-square criterion.

With the savings data there are 7 independent variables. BREG first looks at all possible one predictor equations, and selects the model with the largest R-square. Then BREG examines all two predictor regression models to choose the one with the largest R-square. This continues until all 7 variables are considered.

After the command keyword, BREG, specify the dependent variable, followed by a list of independent variables. Unlike REGRESS and STEPWISE, BREG does not have an argument for the number of predictors. This is because that number keeps changing as BREG proceeds to examining equations of different size models.

179

```
MTB > BREG C1 C2-C8

Best Subsets Regression of Save(%)
```

Vars	R-sq	Adj. R-sq	C-p	s	Income	Asset	Children	Leisure	Loan	Educate	Age
1	48.3	45.8	15.4	9.3634			X				
1	41.2	38.4	20.1	9.9810	X						
2	59.2	55.1	10.2	8.5225			X		X		
2	58.0	53.8	10.9	8.6433			X			X	
3	70.2	65.5	4.8	7.4679			X		X	X	
3	64.5	58.9	8.6	8.1569		X	X			X	
4	74.9	69.3	3.7	7.0459		X	X		X	X	
4	72.9	66.8	5.1	7.3267			X		X	X	X
5	75.8	68.7	5.1	7.1156		X	X		X	X	X
5	75.7	68.5	5.2	7.1340		X	X	X	X	X	
6	77.2	68.7	6.2	7.1198	X	X	X		X	X	X
6	76.1	67.1	6.9	7.2948		X	X	X	X	X	X
7	77.5	66.9	8.0	7.3131	X	X	X	X	X	X	X

BREG prints four statistics for each model: R-sq, adj R-sq, C-p and s. In comparing models with the same size, we can select the one with the largest R-sq. This is the same as choosing the subset with the minimum sum of squares error (SSE). However, as seen in the output, R-square increases with the size of the subset of predictors.

Therefore, while comparing models with unequal numbers of predictors, use the adjusted R-square. The maximum adjusted R-square criterion is analogous to choosing the model with the minimum mean square error (MSE). Using this criterion, from the output we choose the model with 4 predictors, 'Asset', 'Children', 'Loan', and 'Educate' because its adjusted R-square, 69.3, is the largest.

The third criterion in subset selection is the C-p statistic. Choose a model with C-p which is small and close to P, where P represents the number of parameters in the model. P is usually the sum of the number of predictors plus 1, for the intercept. When the model fits the data well, the value of C-p is close to P. A small value of C-p indicates small variance in estimating the true regression coefficients.

The model with the four predictors of 'Asset', 'Children', 'Loan' and 'Educate' has the smallest C-p value of 3.7. However, the model with the four predictors of 'Children', 'Loan', 'Educate' and 'Age' has a C-p value, 5.1, which is closest to the number of parameters, 4 predictors + 1 intercept. This value is also relatively small. In this sense the latter model may be said to satisfy the conditions of C-p best.

Combining the results of using different criteria, adj. R-sq and C-p, it appears that the best model includes 4 predictors: 'Asset', 'Children', 'Loan', and 'Educate'.

10.5 POLYNOMIAL REGRESSION

Polynomial

So far we have assumed linearity in our regression models. Very often, however, the relationship between independent and dependent variables is curvilinear. For example, imagine the amount of work accomplished by a data entry clerk sitting in front of a computer keyboard. Is the work completed a linear function of time? As time progresses, the fatigue factor begins to weigh heavily on work performance.

The following is the data collected at a data entry work station with 8 clerks working at computers. The average number of pages finished by them are computed at every 30-minute interval for 5 hours consecutively.

If the relationship is plotted, it will show a curve, which suggests a polynomial regression. Can this relationship be transformed into a linear regression model?

Exhibit 10.2

# of hours working	# of pages completed
0.5	3.5
1.0	4.2
1.5	4.75
2.0	5.3
2.5	5.8
3.0	5.9
3.5	5.975
4.0	5.9
4.5	5.7
5.0	5.5

Minitab Solution

Let us first plot these two variables. As the following scatterplot indicates, the amount of work completed increases in the beginning, but after a certain point it begins to decline. Also the line is not straight but resembles a parabola.

181

```
MTB > PLOT C2 vs C1;
SUBC> YLABEL 'Work done';
SUBC> XLABEL 'Hours worked'.
```

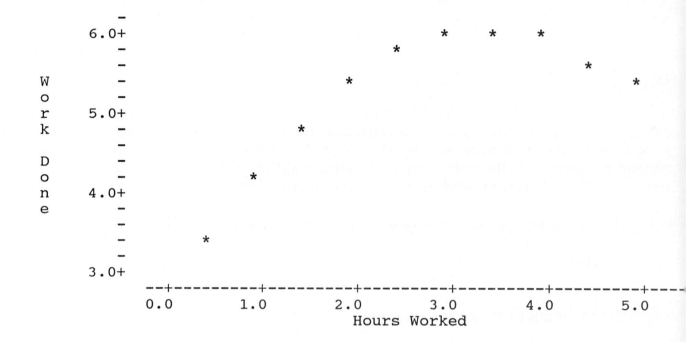

If a linear regression model is used, regressing on the hours worked (C1), the model explains 58.5% of the variation in the amount of work completed. The model is significant at the 0.006 level.

```
MTB > NAME C1 'Hours'  C2 'Work'
MTB > REGRESS C2 1 C1

The regression equation is
Work = 4.04 + 0.441 Hours

Predictor         Coef        Stdev      t-ratio         p
Constant        4.0400       0.3699        10.92     0.000
Hours           0.4409       0.1192         3.70     0.006

s = 0.5415      R-sq = 63.1%     R-sq(adj) = 58.5%

Analysis of Variance

SOURCE          DF           SS           MS          F          p
Regression       1       4.0095       4.0095      13.67      0.006
Error            8       2.3460       0.2933
Total            9       6.3556
```

Although the above model may seem powerful enough at first glance, a much better fit could be achieved using the following model. Now, let us add another term, the quadratic function of 'Hours' into the equation: $Y = a + b_1X + b_2X^2$

A new variable, 'Hour_sq' is created by squaring 'Hours', and stored in C3. 'Work' is regressed on two variables: 'Hours' and 'Hour_sq'. As shown in the output below, this model explains as much as 99.3% of the variation in the work performance.

```
MTB > LET C3=C1**2
MTB > NAME C3 'Hour_sq'
MTB > REGRESS C2 2 C1 C3
```

The regression equation is
Work = 2.58 + 1.90 Hours - 0.265 Hour_sq

Predictor	Coef	Stdev	t-ratio	p
Constant	2.58375	0.08026	32.19	0.000
Hours	1.89716	0.06704	28.30	0.000
Hour_sq	-0.26477	0.01188	-22.29	0.000

s = 0.06824   R-sq = 99.5%   R-sq(adj) = 99.3%

Analysis of Variance

SOURCE	DF	SS	MS	F	p
Regression	2	6.3230	3.1615	678.97	0.000
Error	7	0.0326	0.0047		
Total	9	6.3556			

SOURCE	DF	SEQ SS
C1	1	4.0095
C3	1	2.3135

Unusual Observations

Obs.	C1	C2	Fit	Stdev.Fit	Residual	St.Resid
5	2.50	5.8000	5.6718	0.0323	0.1282	2.13R

R denotes an obs. with a large st. resid.

10.6 MODELS WITH DUMMY VARIABLES

Problem

Not all the independent variables in multiple regression equations need be quantitative. Some could be categorical variables. Although these categorical variables

183

cannot be entered directly in the equation, dummy variables which are created from these, can be included in the model.

Thus far we have not discussed interactive effects within the context of the regression model. However, the question of interactive effects must be addressed whenever there are more than one independent variable in the analysis. In the following example dummy variables are constructed and their interactions are also considered.

Are infant mortality rates lower in states where there are more physicians? Does the relationship vary according to the region? In some areas the practice of midwifery may be more prevalent than elsewhere, which might affect infant mortality rate differently. To answer these questions, data reported in the Statistical Abstract, 1990 are extracted and presented in Exhibit 10.3. The table shows 3 variables. The infant mortality rate represents deaths of infants under one year old per 1,000 live births. Physicians are counted per 100,000 population in each state.

Exhibit 10.3

	Infant mortality	Physicians per 100,000
New England		
Maine	8.3	173
New Hampshire	7.8	186
Vermont	8.5	244
Massachusetts	7.2	322
Rhode Island	8.4	244
Connecticut	8.8	293
Pacific		
Washington	9.7	206
Oregon	10.4	196
California	9.0	242
Alaska	10.4	138
Hawaii	8.9	225
South Atlantic		
Delaware	11.7	191
Maryland	11.5	325
D.C.	19.3	591
Virginia	10.2	204
W.Virginia	9.8	164
N.Carolina	11.9	179
S.Carolina	12.7	156
Georgia	12.7	167
Florida	10.6	203

184

The third variable of 'States' is a categorical one. First, different states are grouped into regions. Then, dummy variables are created to represent three regions as follows:

	Dum1	Dum2
New England	1	0
Pacific	0	1
South Atlantic	0	0

Minitab Solution

After reading infant mortality rates and the number of physicians into C1 and C2, dummy variables are generated by using multipliers in the SET commands. The interaction terms are created by multiplying 'Doctor' and one of the dummy variables, Dum1, Dum2. Regression analysis is performed on the five predictor variables: 'Doctor', two dummy variables(Dum1, Dum2), and two interaction terms(Int1, Int2).

```
MTB > READ C1-C2
DATA> 8.3   173
DATA> ...
DATA> END
MTB > SET C3
DATA> 6(1) 5(0) 9(0)
DATA> END
MTB > SET C4
DATA> 6(0) 5(1) 9(0)
DATA> END
MTB > LET C7=C2*C3
MTB > LET C8=C2*C4
MTB > NAME C1 'Mortal'  C2 'Doctor'
MTB > NAME C3 'Dum1'  C4 'Dum2'
MTB > NAME C7 'Int1' C8 'Int2'
MTB > REGRESS C1 5 C2 C3 C4 C7 C8
```

The regression equation is
Mortal = 8.10 + 0.0172 Doctor + 0.54 Dum1 + 4.65 Dum2
         - 0.0191 Int1 - 0.0324 Int2

Predictor	Coef	Stdev	t-ratio	p
Constant	8.0957	0.8235	9.83	0.000
Doctor	0.017220	0.002984	5.77	0.000
Dum1	0.541	2.416	0.22	0.826
Dum2	4.648	3.165	1.47	0.164
Int1	-0.019148	0.009585	-2.00	0.066
Int2	-0.03243	0.01524	-2.13	0.052

s = 1.184   R-sq = 85.1%   R-sq(adj) = 79.8%

```
Analysis of Variance

SOURCE          DF          SS          MS          F          p
Regression       5      112.043      22.409      15.99      0.000
Error           14       19.615       1.401
Total           19      131.658

SOURCE          DF      SEQ SS
Doctor           1      39.195
Dum1             1      49.070
Dum2             1      12.523
Int1             1       4.907
Int2             1       6.347

Unusual Observations
Obs.      C2          C1          Fit  Stdev.Fit  Residual   St.Resid
 13      325      11.500      13.692      0.466     -2.192      -2.01R
 14      591      19.300      18.273      1.113      1.027       2.55R

      R denotes an obs. with a large st. resid.
```

The regression equation is given as:

$$\text{Mortal} = 8.10 + 0.0172 \text{ Doctor} + 0.54 \text{ Dum1} + 4.65 \text{ Dum2} - 0.0191 \text{ Int1} - 0.0324 \text{ Int2}$$

The models specific to the regions are as follows:

| New England: | Mortal | $= 8.10 + 0.54 + (0.0172 - 0.0191) \text{ Doctor}$ |
| | | $= 8.64 - 0.0019 \text{ Doctor}$ |

| Pacific: | Mortal | $= 8.10 + 4.65 + (0.0172 - 0.0324) \text{ Doctor}$ |
| | | $= 12.75 - 0.0152 \text{ Doctor}$ |

| South Atlantic: | Mortal | $= 8.10 + 0.0172 \text{ Doctor}$ |

This model, which is significant at the 0.000(=.0005) level, explains 79.8% of the variation in the infant mortality rates.

However, the model contains interaction effects which are not significant at the 0.05 level. When interaction terms are included, it is difficult to interpret the main effects. Therefore it makes sense to eliminate nonsignificant interaction terms, and run the regression analysis again. A portion of the results is shown below:

```
MTB > REGRESS 'Mortal' 3 'Doctor' 'Dum1' 'Dum2'

  The regression equation is
  Mortal = 8.80 + 0.0143 Doctor - 4.12 Dum1 - 2.00 Dum2
```

Predictor	Coef	Stdev	t-ratio	p
Constant	8.8028	0.9174	9.60	0.000
Doctor	0.014300	0.003269	4.37	0.000
Dum1	−4.1207	0.7321	−5.63	0.000
Dum2	−2.0029	0.7862	−2.55	0.022

s = 1.389   R-sq = 76.6%   R-sq(adj) = 72.2%

This model without interaction terms is significant at the 0.000(=0.0005) level, and explains 72.2% of the variation in infant mortality rates. The regression equation of the whole model is decomposed for the three regions as follows:

New England:	Mortal = 4.68 + 0.0143 Doctor
Pacific:	Mortal = 6.80 + 0.0143 Doctor
South Atlantic:	Mortal = 8.80 + 0.0143 Doctor

The constant(8.8028) in the overall regression is significant at the 0.000(=.0005) level. This intercept refers to the infant mortality rate without any physicians available in the South Atlantic region, which is used as an arbitrary baseline category. The coefficient of 'Dum1' is significant at the 0.000(=.0005) level, which reflects the partial effect of region, that is, the expected infant mortality difference between the New England and the South Atlantic regions. The difference between the Pacific and the South Atlantic regions, represented by 'Dum2', is also significant at the 0.022 level. Therefore we can reject the null hypothesis that regional differences in infant mortality rates are zero. Finally, the effect of the physician supply upon infant mortality rates is significant at the 0.000(=0.0005) level.

10.7 MORE ON COMMANDS

10.7.1 The STEPWISE Command

Syntax:

```
STEPWISE regression of C on the predictors C,...,C

     FENTER = K                          (default is 4)
     FREMOVE = K                         (default is 4)
     FORCE C,...,C
     ENTER C,...,C
     REMOVE C,...,C
     BEST K alternative predictors       (default is 0)
     STEPS = K                           (default depends on
                                          output width)
```

The STEPWISE command performs a stepwise regression analysis with three methods: stepwise, forward selection, and backward elimination. In addition, you can intervene in the analysis at the point you select.

187

Forward Selection: At each step the independent variable which best meets the criterion of FENTER (see below) is entered into the equation. No predictor, once added to the model, is to be removed in this method. Therefore set FREMOVE = 0.

Backward Elimination: The model starts with its full set of predictors listed in the command. Then, at each step the independent variable which does not meet the criterion of FREMOVE (see below) is removed. No predictor is allowed to reenter the equation. Therefore, set FENTER = 100000. Also all predictors must be listed on the ENTER subcommand.

Stepwise Procedure: This is the default, and is a combination of the above two methods.

Interactive Resetting: The STEPWISE command proceeds automatically for one "page" without interruption. The number of steps to be contained in a page can be set by subcommand STEPS. At the end of a page, you are asked if you wish to use any subcommand. You may then type some subcommands, such as changing the value of FENTER, or REMOVing some of the predictors in the equation. STEPWISE then continues automatically for another page.

The subcommands are explained briefly below:

FENTER=K   Set the F criterion value, K, for inclusion. Variables with F greater than K will be considered for inclusion into the equation. The default value of K is 4.

FREMOVE=K   Set the F criterion value, K, for removal. Among the variables with F smaller than K, the one with the smallest F is removed from the equation. The default is 4.

FORCE C,...,C   Specify columns containing the variables which are not removed by the automatic stepwise procedure even if their F's fall below FREMOVE.

ENTER C,...,C   Specify variables to be entered into the equation. They can be removed if their F's fall below FREMOVE.

REMOVE C,...,C   Specify variables to be removed from the equation, any of which may be reentered later if its F statistic is large enough.

BEST K alternative variables
            This specifies the number(K) of best alternative variables to be printed at any step where a variable is added to the equation. The default is 0.

STEPS = K   This specifies the number of steps per page before the MORE? prompt appears so that you can interrupt the process.

10.7.2 The BREG Command

Syntax:
```
BREGRESS C on predictors C,...,C

     INCLUDE    BEST    NVARS    NOCONSTANT
```

The subcommands are explained below:

INCLUDE C,...,C   This subcommands specifies columns to be included as predictors in all of the subset regression models. These columns should come from those already listed under the BREG command.

BEST K   This subcommand prints information from the best K models of each size. Up to 5 of the best model of each size can be printed. The default is 2.

NVARS K [up to K]   This subcommand specifies the starting and ending number of predictors for subsets printout. For example, NVARS 3 10 will print only the best 3, 4, ..., 10 predictor models.


```
NEW MINITAB COMMANDS

    NSCORE
    STEPWISE
    BREG
```


EXERCISE

1.   Calculate the regression equation for the dependent variable, GPA, using the number of study hours, the number of work hours, the number of units carried, the class level, and marital status. Create dummy variables for marital status.

2.   Simplify the model into a more parsimonious one, using STEPWISE and BREG.

INDEX

ABSOLUTE	45	DELETE	28-29,37
ADD	42	DESCRIBE	49-50
ANOVA	132-133	DIVIDE	43
AOVONEWAY	120-121	DOS commands CD,DIR,TYPE	60-61
arithmetic function	44	DOTPLOT	67,122
arithmetic operation	42-43	ECHO	98,101-102,111
BASE	98,100-101,111	END	7,98,101-102
BOXPLOT	69-71	ERASE	29
BREG	179-181,189	EXECUTE	62-63,98, 101-102,111
BRIEF	135,164,168	File type	63
CDF Binomial Chisquare Normal	88-89 153 89,91	FRIEDMAN	147-148
		HELP	10
CHISQUARE	152-155	HEIGHT	74
CODE	34,48	HISTOGRAM	64-66
col-wise functions	45	INDICATOR	85-86,135
col-wise statistics	45-47	INFORMATION	26-27
		INSERT	27-28,36
CONCATENATE	56-57	INVCDF Binomial Chisquare F Normal T	89 153 113,127, 129-130,136 89 178
CONVERT	50,55-56,66		
COPY	50-52		
CORRELATE	156,159,177		
COUNT	46,73,113	JOURNAL	60
Ctrl-Brk	11	KRUSKAL-WALLIS	149
Data Editor	17-25	LET	9-10,40-42
		LPLOT	76-77,131

Macro	97-98,101, 110-111	Integer	82,92-93,96-97, 100-101
MANN-WHITNEY	145-146	Normal	100
MEAN	46	RANK	31-32,156
missing data	13	READ	8,13-14
		from file	65
MOOD	150-151	REGRESS	134-136,163-170, 173-175,182-183, 185-187
MPLOT	74-75		
MULTIPLY	43	RESTART	29
NAME	8-9,14	RETRIEVE	25-26,36
NOECHO	98,101-102,111	RMEAN	47
NOJOURNAL	60,63	ROUND	44,94
NOOUTFILE	59,63	row-wise statistics	47
NOPAPER	59		
NOTE(#)	9	RUNS	139
NSCORES	176	SAMPLE	92-95,101,111
ONEWAY	121-122	SAVE	25,34-35
OUTFILE	58-59	SET	6-7,11-12
PAPER	59	SINTERVAL	141,142
PARSUMS	83-84	SORT	29-31,37
PDF		SQRT	47
Binomial	87-88	STACK	52-54,115-116, 122
Integer	95		
Normal	91		
Poisson	89-90	STDEV	47
PLOT	71-74	STEM-AND-LEAF	68-69
PRINT	7,8,15	STEPWISE	177-179,187-189
RAISE	47	STEST	140-141
RANDOM		STOP	10
Bernoulli	81,83,92		
Binomial	100-101	STORE	98,101-102,111
Discrete	84-85,93		

SUBTRACT	43,47
SUM	46
SYSTEM	61-62
TABLE	32-33,37-38,48
TALLY	33,38
TINTERVAL	108,111,117
TPLOT	77-79
TTEST	109,117-118
Tukey	123-124
TWOSAMPLE	112-115
TWOT	115-116
TWOWAY	125-126,128-129
UNSTACK	54-55,94-95
WIDTH	74
WINTERVAL	144
worksheet	2-3
WRITE	27,35
WTEST	143
ZINTERVAL	105
ZTEST	106-107